THE INTERNATIONAL SCIENTIFIC SERIES.

VOL VII.

THE INTERNATIONAL SCIENTIFIC SERIES.

Works already Published.

THE INTERNATIONAL SCIENTIFIC SERIES.

THE CONSERVATION OF ENERGY.

BY

BALFOUR STEWART, LL. D., F. R. S.,

PROFESSOR OF NATURAL PHILOSOPHY AT THE OWENS COLLEGE, MANCHESTER.

WITH AN APPENDIX,

TREATING OF THE VITAL AND MENTAL APPLICATIONS OF THE DOCTRINE.

NEW YORK:
D. APPLETON AND COMPANY,
549 & 551 BROADWAY.
1875.

NOTE TO THE AMERICAN EDITION.

THE great prominence which the modern doctrine of the Conservation of Energy or Correlation of Forces has lately assumed in the world of thought, has made a simple and popular explanation of the subject very desirable. The present work of Dr. Balfour Stewart, contributed to the International Scientific Series, fully meets this requirement, as it is probably the clearest and most elementary statement of the question that has yet been attempted. Simple in language, copious and familiar in illustration, and remarkably lucid in the presentation of facts and principles, his little treatise forms just the introduction to the great problem of the interaction of natural forces that is required by general readers. But Professor Stewart having confined himself mainly to the physical aspects of the subject, it was desirable that his views should be supplemented by a statement of the operation of the principle in the spheres of life and mind. An Appendix has, accordingly, been added to the American edition of Dr. Stewart's work, in which these applications of the law are considered.

Professor Joseph Le Conte published a very able essay fourteen years ago on the Correlation of the Physical and Vital Forces, which was extensively reprinted abroad, and placed the name of the author among the leading interpreters of the subject. His mode of presenting it was regarded as peculiarly happy, and was widely adopted by other writers. After further investigations and more mature reflection, he has recently restated his views, and has kindly furnished the revised essay for insertion in this volume.

Professor A. Bain, the celebrated Psychologist of Aberdeen, who has done so much to advance the study of mind in its physiological relations, prepared an interesting lecture not long ago on the "Correlation of the Nervous and Mental Forces," which was read with much interest at the time of its publication, and is now reprinted as a suitable exposition of that branch of the subject. These two essays, by carrying out the principle in the field of vital and mental phenomena, will serve to give completeness and much greater value to the present volume.

New York, *December*, 1873.

PREFACE.

We may regard the Universe in the light of a vast physical machine, and our knowledge of it may be conveniently divided into two branches.

The one of these embraces what we know regarding the structure of the machine itself, and the other what we know regarding its method of working.

It has appeared to the author that, in a treatise like this, these two branches of knowledge ought as much as possible to be studied together, and he has therefore endeavored to adopt this course in the following pages. He has regarded a universe composed of atoms with some sort of medium between them as the machine, and the laws of energy as the laws of working of this machine.

The first chapter embraces what we know regarding atoms, and gives also a definition of Energy. The various forces and energies of Nature are thereafter enumerated, and the law of Conservation is stated. Then follow the various transmutations of Energy, according to a list, for which the author is indebted to Prof. Tait. The fifth chapter gives a short historical sketch of the subject, ending with the law of Dissipation; while the sixth and last chapter gives some account of the position of living beings in this universe of Energy. B. S.

The Owens College, Manchester,
August, 1873.

CONTENTS.

THE CONSERVATION OF ENERGY.

CHAPTER I.

WHAT IS ENERGY?

Our Ignorance of Individuals.

1. VERY often we know little or nothing of individuals, while we yet possess a definite knowledge of the laws which regulate communities.

The Registrar-General, for example, will tell us that the death-rate in London varies with the temperature in such a manner that a very low temperature is invariably accompanied by a very high death-rate. But if we ask him to select some one individual, and explain to us in what manner his death was caused by the low temperature, he will, most probably, be unable to do so.

Again, we may be quite sure that after a bad harvest there will be a large importation of wheat into the country, while, at the same time, we are quite ignorant of the individual journeys of the various particles of flour that go to make up a loaf of bread.

Or yet again, we know that there is a constant carriage of air from the poles to the equator, as shown by the trade winds, and yet no man is able to individualize a particle of this air, and describe its various motions.

2. Nor is our knowledge of individuals greater in the domains of physical science. We know nothing, or next to nothing, of the ultimate structure and properties of matter, whether organic or inorganic.

No doubt there are certain cases where a large number of particles are linked together, so as to act as one individual, and then we can predict its action—as, for instance, in the solar system, where the physical astronomer is able to foretell with great exactness the positions of the various planets, or of the moon. And so, in human affairs, we find a large number of individuals acting together as one nation, and the sagacious statesman taking very much the place of the sagacious astronomer, with regard to the action and reaction of various nations upon one another.

But if we ask the astronomer or the statesman to select an individual particle and an individual human being, and predict the motions of each, we shall find that both will be completely at fault.

3. Nor have we far to look for the cause of their ignorance. A continuous and restless, nay, a very complicated, activity is the order of nature throughout all her indi-

viduals, whether these be living beings or inanimate particles of matter. Existence is, in truth, one continued fight, and a great battle is always and everywhere raging, although the field in which it is fought is often completely shrouded from our view.

4. Nevertheless, although we cannot trace the motions of individuals, we may sometimes tell the result of the fight, and even predict how the day will go, as well as specify the causes that contribute to bring about the issue.

With great freedom of action and much complication of motion in the individual, there are yet comparatively simple laws regulating the joint result attainable by the community.

But, before proceeding to these, it may not be out of place to take a very brief survey of the organic and inorganic worlds, in order that our readers, as well as ourselves, may realize our common ignorance of the ultimate structure and properties of matter.

5. Let us begin by referring to the causes which bring about disease. It is only very recently that we have begun to suspect a large number of our diseases to be caused by organic germs. Now, assuming that we are right in this, it must nevertheless be confessed that our ignorance about these germs is most complete. It is perhaps doubtful whether we ever saw one of these organisms,*

* It is said that there are one or two instances where the microscope has enlarged them into visibility.

while it is certain that we are in profound ignorance of their properties and habits.

We are told by some writers * that the very air we breathe is absolutely teeming with germs, and that we are surrounded on all sides by an innumerable array of minute organic beings. It has also been conjectured that they are at incessant warfare among themselves, and that we form the spoil of the stronger party. Be this as it may, we are at any rate intimately bound up with, and, so to speak, at the mercy of, a world of creatures, of which we know as little as of the inhabitants of the planet Mars.

6. Yet, even here, with profound ignorance of the individual, we are not altogether unacquainted with some of the habits of these powerful predatory communities. Thus we know that cholera is eminently a low level disease, and that during its ravages we ought to pay particular attention to the water we drink. This is a general law of cholera, which is of the more importance to us because we cannot study the habits of the individual organisms that cause the disease.

Could we but see these, and experiment upon them, we should soon acquire a much more extensive knowledge of their habits, and perhaps find out the means of extirpating the disease, and of preventing its recurrence.

Again, we know (thanks to Jenner) that vaccination will prevent the ravages of small-pox, but in this in-

* *See* Dr. Angus Smith on Air and Rain.

stance we are no better off than a band of captives who have found out in what manner to mutilate themselves, so as to render them uninteresting to their victorious foe.

7. But if our knowledge of the nature and habits of organized molecules be so small, our knowledge of the ultimate molecules of inorganic matter is, if possible, still smaller. It is only very recently that the leading men of science have come to consider their very existence as a settled point.

In order to realize what is meant by an inorganic molecule, let us take some sand and grind it into smaller and smaller particles, and these again into still smaller. In point of fact we shall never reach the superlative degree of smallness by this operation—yet in our imagination we may suppose the sub-division to be carried on continuously, always making the particles smaller and smaller. In this case we should, at last, come to an ultimate molecule of sand or oxide of silicon, or, in other words, we should arrive at the smallest entity retaining all the properties of sand, so that were it possible to divide the molecule further the only result would be to separate it into its chemical constituents, consisting of silicon on the one side and oxygen on the other.

We have, in truth, much reason to believe that sand, or any other substance, is incapable of infinite subdivision, and that all we can do in grinding down a solid lump of anything is to reduce it into lumps similar to the original, but only less in size, each of these small lumps containing probably a great number of individual molecules.

8. Now, a drop of water no less than a grain of sand is built up of a very great number of molecules, attached to one another by the force of cohesion—a force which is much stronger in the sand than in the water, but which nevertheless exists in both. And, moreover, Sir William Thomson, the distinguished physicist, has recently arrived at the following conclusion with regard to the size of the molecules of water. He imagines a single drop of water to be magnified until it becomes as large as the earth, having a diameter of 8000 miles, and all the molecules to be magnified in the same proportion; and he then concludes that a single molecule will appear, under these circumstances, as somewhat larger than a shot, and somewhat smaller than a cricket ball.

9. Whatever be the value of this conclusion, it enables us to realize the exceedingly small size of the individual molecules of matter, and renders it quite certain that we shall never, by means of the most powerful microscope, succeed in making visible these ultimate molecules. For our knowledge of the sizes, shapes, and properties of such bodies, we must always, therefore, be indebted to indirect evidence of a very complicated nature.

It thus appears that we know little or nothing about the shape or size of molecules, or about the forces which actuate them; and, moreover, the very largest masses of the universe share with the very smallest this property of being beyond the direct scrutiny of the human senses —the one set because they are so far away, and the other because they are so small.

10. Again, these molecules are not at rest, but, on the contrary, they display an intense and ceaseless energy in their motions. There is, indeed, an uninterrupted warfare going on—a constant clashing together of these minute bodies, which are continually maimed, and yet always recover themselves, until, perhaps, some blow is struck sufficiently powerful to dissever the two or more simple atoms that go to form a compound molecule. A new state of things thenceforward is the result.

But a simple elementary atom is truly an immortal being, and enjoys the privilege of remaining unaltered and essentially unaffected amid the most powerful blows that can be dealt against it—it is probably in a state of ceaseless activity and change of form, but it is nevertheless always the same.

11. Now, a little reflection will convince us that we have in this ceaseless activity another barrier to an intimate acquaintance with molecules and atoms, for even if we could see them they would not remain at rest sufficiently long to enable us to scrutinize them.

No doubt there are devices by means of which we can render visible, for instance, the pattern of a quickly revolving coloured disc, for we may illuminate it by a flash of electricity, and the disc may be supposed to be stationary during the extremely short time of the flash.

But we cannot say the same about molecules and atoms, for, could we see an atom, and could we illuminate it by a flash of electricity, the atom would most probably have vibrated many times during the exceedingly small time of the flash. In fine, the limits placed upon our senses, with respect to space and time, equally preclude the possibility of our ever becoming directly acquainted with these exceedingly minute bodies, which are nevertheless the raw materials of which the whole universe is built.

Action and Reaction, Equal and Opposite.

12. But while an impenetrable veil is drawn over the individual in this warfare of clashing atoms, yet we are not left in profound ignorance of the laws which determine the ultimate result of all these motions, taken together as a whole.

In a Vessel of Goldfish.

Let us suppose, for instance, that we have a glass globe containing numerous goldfish standing on the table, and delicately poised on wheels, so that the slightest push, the one way or the other, would make it move. These goldfish are in active and irregular motion, and he would be a very bold man who should venture to predict the movements of an individual fish. But of one thing we may be quite certain: we may rest assured that, notwithstanding all the irregular motions of its living inhabitants, the globe containing the goldfish will remain at rest upon its wheels.

Even if the table were a lake of ice, and the wheels were extremely delicate, we should find that the globe would remain at rest. Indeed, we should be exceedingly surprised if we found the globe going away of its own accord from the one side of the table to the other, or from the one side of a sheet of ice to the other, in consequence of the internal motions of its inhabitants. Whatever be the motions of these individual units, yet we feel sure that the globe cannot move itself *as a whole.* In such a system, therefore, and, indeed, in every system left to itself, there may be strong internal forces acting between the various parts, but these *actions and reactions are equal and opposite,* so that while the small parts, whether visible or invisible, are in violent commotion among themselves, yet the system as a whole will remain at rest.

In a Rifle.

13. Now it is quite a legitimate step to pass from this instance of the goldfish to that of a rifle that has just been fired. In the former case, we imagined the globe, together with its fishes, to form one system; and in the latter, we must look upon the rifle, with its powder and ball, as forming one system also.

Let us suppose that the explosion takes place through the application of a spark. Although this spark is an external agent, yet if we reflect a little we shall see that its only office in this case is to summon up the internal forces already existing in the loaded rifle, and bring them into vigorous action, and that in virtue of these internal forces the explosion takes place.

The most prominent result of this explosion is the outrush of the rifle ball with a velocity that may, perhaps, carry it for the best part of a mile before it comes to rest; and here it would seem to us, at first sight, that the law of equal action and reaction is certainly broken, for these internal forces present in the rifle have at least propelled part of the system, namely, the rifle ball, with a most enormous velocity in one direction.

14. But a little further reflection will bring to light another phenomenon besides the out-rush of the ball. It is well known to all sportsmen that when a fowling-piece is discharged, there is a kick or recoil of the piece itself against the shoulder of the sportsman, which he would rather get rid of, but which we most gladly welcome as the solution of our difficulty. In plain terms, while the ball is projected forwards, the rifle stock (if free to move) is at the same moment projected backwards. To fix our ideas, let us suppose that the rifle stock weighs 100 ounces, and the ball one ounce, and that the ball is projected forwards with the velocity of 1000 feet per second; then it is asserted, by the law of action and reaction, that the rifle stock is at the same time projected backwards with the velocity of 10 feet per second, so that the mass of the stock, multiplied by its velocity of

recoil, shall precisely equal the mass of the ball, multiplied by its velocity of projection. The one product forms a measure of the action in the one direction, and the other of the reaction in the opposite direction, and thus we see that in the case of a rifle, as well as in that of the globe of fish, action and reaction are equal and opposite.

In a Falling Stone.

15. We may even extend the law to cases in which we do not perceive the recoil or reaction at all. Thus, if I drop a stone from the top of a precipice to the earth, the motion seems all to be in one direction, while at the same time it is in truth the result of a mutual attraction between the earth and the stone. Does not the earth move also? We cannot see it move, but we are entitled to assert that it does in reality move upwards to meet the stone, although quite to an imperceptible extent, and that the law of action and reaction holds here as truly as in a rifle, the only difference being that in the one case the two objects are rushing together, while in the other they are rushing apart. Inasmuch, however, as the mass of the earth is very great compared with that of the stone, it follows that its velocity must be extremely small, in order that the mass of the earth, multiplied into its velocity upwards, shall equal the mass of the stone, multiplied into its velocity downwards.

16. We have thus, in spite of our ignorance of the ultimate atoms and molecules of matter, arrived at a

general law which regulates the action of internal forces. We see that these forces are always mutually exerted, and that if A attracts or repels B, B in its turn attracts or repels A. We have here, in fact, a very good instance of that kind of generalization, which we may arrive at, even in spite of our ignorance of individuals.

But having now arrived at this law of action and reaction, do we know all that it is desirable to know? have we got a complete understanding of what takes place in all such cases—for instance, in that of the rifle which is just discharged? Let us consider this point a little further.

The Rifle further considered.

17. We define quantity of motion to mean the product of the mass by the velocity; and since the velocity of recoil of the rifle stock, multiplied by the mass of the stock, is equal to the velocity of projection of the rifle ball, multiplied by the mass of the ball, we conceive ourselves entitled to say that the quantity of motion, or momentum, generated is equal in both directions, so that the law of action and reaction holds here also. Nevertheless, it cannot but occur to us that, *in some sense*, the motion of the rifle ball is a very different thing from that of the stock, for it is one thing to allow the stock to recoil against your shoulder and discharge the ball into the air, and a very different thing to discharge the ball against your shoulder and allow the stock to fly into the

air. And if any man should assert the absolute equality between the blow of the rifle stock and that of the rifle ball, you might request him to put his assertion to this practical test, with the absolute certainty that he would decline. Equality between the two!—Impossible! Why, if this were the case, a company of soldiers engaged in war would suffer much more than the enemy against whom they fired, for the soldiers would certainly feel each recoil, while the enemy would suffer from only a small proportion of the bullets.

The Rifle Ball possesses Energy.

18. Now, what is the meaning of this great difference between the two? We have a vivid perception of a mighty difference, and it only remains for us to clothe our naked impressions in a properly fitting scientific garb.

The something which the rifle ball possesses in contradistinction to the rifle stock is clearly the power of overcoming resistance. It can penetrate through oak wood or through water, or (alas! that it should be so often tried) through the human body, and this power of penetration is the distinguishing characteristic of a substance moving with very great velocity.

19. Let us define by the term *energy* this power which the rifle ball possesses of overcoming obstacles or of doing work. Of course we use the word work without reference to the moral character of the thing done, and con-

ceive ourselves entitled to sum up, with perfect propriety and innocence, the amount of work done in drilling a hole through a deal board or through a man.

20. A body such as a rifle ball, moving with very great velocity, has, therefore, energy, and it requires very little consideration to perceive that this *energy will be proportional to its weight or mass,* for a ball of two ounces moving with the velocity of 1000 feet per second will be the same as two balls of one ounce moving with this velocity, but the energy of two similarly moving ounce balls will manifestly be double that of one, so that the energy is proportional to the weight, if we imagine that, meanwhile, the velocity remains the same.

21. But, on the other hand, the energy is not simply proportional to the velocity, for, if it were, the energy of the rifle stock and of the rifle ball would be the same, inasmuch as the rifle stock would gain as much by its superior mass as it would lose by its inferior velocity. Therefore, the energy of a moving body increases with the velocity more quickly than a simple proportion, so that if the velocity be doubled, the energy is more than doubled. Now, in what manner does the energy increase with the velocity? That is the question we have now to answer, and, in doing so, we must appeal to the familiar facts of everyday observation and experience.

22. In the first place, it is well known to artillerymen, that if a ball have a double velocity, its penetrating power or energy is increased nearly fourfold, so that it

will pierce through four, or nearly four, times as many deal boards as the ball with only a single velocity—in other words, they will tell us, in mathematical language, that the energy varies as the square of the velocity.

Definition of Work.

23. And now, before proceeding further, it will be necessary to tell our readers how to measure work in a strictly scientific manner. We have defined energy to be the power of doing work, and although every one has a general notion of what is meant by work, that notion may not be sufficiently precise for the purpose of this volume. How, then, are we to measure work? Fortunately, we have not far to go for a practical means of doing this. Indeed, there is a force at hand which enables us to accomplish this measurement with the greatest precision, and this force is gravity. Now, the first operation in any kind of numerical estimate is to fix upon our unit or standard. Thus we say a rod is so many inches long, or a road so many miles long. Here an inch and a mile are chosen as our standards. In like manner, we speak of so many seconds, or minutes, or hours, or days, or years, choosing that standard of time or duration which is most convenient for our purpose. So in like manner we must choose our unit of work, but in order to do so we must first of all choose our units of weight and of length, and for these we will take the *kilogramme* and the *metre*, these being the units of the metrical system. The kilo-

gramme corresponds to about 15,432·35 English grains, being rather more than two pounds avoirdupois, and the metre to about 39·371 English inches.

Now, if we raise a kilogramme weight one metre in vertical height, we are conscious of putting forth an effort to do so, and of being resisted in the act by the force of gravity. In other words, we spend energy and do work in the process of raising this weight.

Let us agree to consider the energy spent, or the work done, in this operation as one unit of work, and let us call it the *kilogrammetre.*

24. In the next place, it is very obvious that if we raise the kilogramme two metres in height, we do two units of work—if three metres, three units, and so on.

And again, it is equally obvious that if we raise a weight of two kilogrammes one metre high, we likewise do two units of work, while if we raise it two metres high, we do four units, and so on.

From these examples we are entitled to derive the following rule:—*Multiply the weight raised (in kilogrammes) by the vertical height (in metres) through which it is raised, and the result will be the work done (in kilogrammetres).*

Relation between Velocity and Energy.

25. Having thus laid a numerical foundation for our superstructure, let us next proceed to investigate the relation between velocity and energy. But first let us say a

few words about velocity. This is one of the few cases in which everyday experience will aid, rather than hinder, us in our scientific conception. Indeed, we have constantly before us the example of bodies moving with variable velocities.

Thus a railway train is approaching a station and is just beginning to slacken its pace. When we begin to observe, it is moving at the rate of forty miles an hour. A minute afterwards it is moving at the rate of twenty miles only, and a minute after that it is at rest. For no two consecutive moments has this train continued to move at the same rate, and yet we may say, with perfect propriety, that at such a moment the train was moving, say, at the rate of thirty miles an hour. We mean, of course, that had it continued to move for an hour with the speed which it had when we made the observation, it would have gone over thirty miles. We know that, as a matter of fact, it did not move for two seconds at that rate, but this is of no consequence, and hardly at all interferes with our mental grasp of the problem, so accustomed are we all to cases of variable velocity.

26. Let us now imagine a kilogramme weight to be shot vertically upwards, with a certain initial velocity—let us say, with the velocity of 9·8 metres in one second. Gravity will, of course, act against the weight, and continually diminish its upward speed, just as in the railway train the break was constantly reducing the

velocity. But yet it is very easy to see what is meant by an initial velocity of 9 · 8 metres per second; it means that if gravity did not interfere, and if the air did not resist, and, in fine, if no external influence of any kind were allowed to act upon the ascending mass, it would be found to move over 9 · 8 metres in one second.

Now, it is well known to those who have studied the laws of motion, that a body, shot upwards with the velocity of 9 · 8 metres in one second, will be brought to rest when it has risen 4 · 9 metres in height. If, therefore, it be a kilogramme, its upward velocity will have enabled it to raise itself 4 · 9 metres in height against the force of gravity, or, in other words, it will have done 4 · 9 units of work; and we may imagine it, when at the top of its ascent, and just about to turn, caught in the hand and lodged on the top of a house, instead of being allowed to fall again to the ground. We are, therefore, entitled to say that a kilogramme, shot upwards with the velocity of 9 · 8 metres per second, has energy equal to 4 · 9, inasmuch as it can raise itself 4 · 9 metres in height.

27. Let us next suppose that the velocity with which the kilogramme is shot upwards is that of 19 · 6 metres per second. It is known to all who have studied dynamics that the kilogramme will now mount not only twice, but four times as high as it did in the last instance—in other words, it will now mount 19 · 6 metres in height.

Evidently, then, in accordance with our principles of

measurement, the kilogramme has now four times as much energy as it had in the last instance, because it can raise itself four times as high, and therefore do four times as much work, and thus we see that the energy is increased four times by doubling the velocity.

Had the initial velocity been three times that of the first instance, or 29·4 metres per second, it might in like manner be shown that the height attained would have been 44·1 metres, so that by tripling the velocity the energy is increased nine times.

28. We thus see that whether we measure the energy of a moving body by the thickness of the planks through which it can pierce its way, or by the height to which it can raise itself against gravity, the result arrived at is the same. *We find the energy to be proportional to the square of the velocity,* and we may formularize our conclusion as follows:—

Let v = the initial velocity expressed in metres per second, then the energy in kilogrammetres $= \frac{v^2}{19 \cdot 6}$. Of course, if the body shot upwards weighs two kilogrammes, then everything is doubled, if three kilogrammes, tripled, and so on; so that finally, if we denote by m the mass of the body in kilogrammes, we shall have the energy in kilogrammetres $= \frac{m\,v^2}{19 \cdot 6}$. To test the truth of this formula, we have only to apply it to the cases described in Arts. 26 and 27.

29. We may further illustrate it by one or two examples. For instance, let it be required to find the energy contained in a mass of five kilogrammes, shot upwards with the velocity of 20 metres per second.

Here we have $m = 5$ and $v = 20$, hence—

$$\text{Energy} = \frac{5\ (20)^2}{19 \cdot 6} = \frac{2000}{19 \cdot 6} = 102 \cdot 04 \text{ nearly.}$$

Again, let it be required to find the height to which the mass of the last question will ascend before it stops. We know that its energy is 102·04, and that its mass is 5. Dividing 102·04 by 5, we obtain 20·408 as the height to which this mass of five kilogrammes must ascend in order to do work equal to 102·04 kilogrammetres.

30. In what we have said we have taken no account either of the resistance or of the buoyancy of the atmosphere; in fact, we have supposed the experiments to be made in vacuo, or, if not in vacuo, made by means of a heavy mass, like lead, which will be very little influenced either by the resistance or buoyancy of the air.

We must not, however, forget that if a sheet of paper, or a feather, be shot upwards with the velocities mentioned in our text, they will certainly not rise in the air to nearly the height recorded, but will be much sooner brought to a stop by the very great resistance which they encounter from the air, on account of their great surface, combined with their small mass.

On the other hand, if the substance we make use of be a large light bag filled with hydrogen, it will find its way upwards without any effort on our part, and we shall certainly be doing no work by carrying it one or more metres in height—it will, in reality, help to pull us up, instead of requiring help from us to cause it to ascend. In fine, what we have said is meant to refer to the force of gravity alone, without taking into account a resisting medium such as the atmosphere, the existence of which need not be considered in our present calculations.

31. It should likewise be remembered, that while the energy of a moving body depends upon its velocity, it is independent of the direction in which the body is moving. We have supposed the body to be shot upwards with a given velocity, but it might be shot horizontally with the same velocity, when it would have precisely the same energy as before. A cannon ball, if fired vertically upwards, may either be made to spend its energy in raising itself, or in piercing through a series of deal boards. Now, if the same ball be fired horizontally with the same velocity it will pierce through the same number of deal boards.

In fine, direction of motion is of no consequence, and the only reason why we have chosen vertical motion is that, in this case, there is always the force of gravity steadily and constantly opposing the motion of the body, and enabling us to obtain an accurate measure of the work which it does by piercing its way upwards against this force.

32. But gravity is not the only force, and we might

measure the energy of a moving body by the extent to which it would bend a powerful spring or resist the attraction of a powerful magnet, or, in fine, we might make use of the force which best suits our purpose. If this force be a constant one, we must measure the energy of the moving body by the space which it is able to traverse against the action of the force—just as, in the case of gravity, we measured the energy of the body by the space through which it was able to raise itself against its own weight.

33. We must, of course, bear in mind that if this force be more powerful than gravity, a body moved a short distance against it will represent the expenditure of as much energy as if it were moved a greater distance against gravity. In fine, we must take account both of the strength of the force and of the distance moved over by the body against it before we can estimate in an accurate matter the work which has been done.

CHAPTER II.

MECHANICAL ENERGY AND ITS CHANGE INTO HEAT.

Energy of Position. A Stone high up.

34. In the last chapter it was shown what is meant by energy, and how it depends upon the velocity of a moving body; and now let us state that this same energy or power of doing work may nevertheless be possessed by a body absolutely at rest. It will be remembered (Art. 26) that in one case where a kilogramme was shot vertically upwards, we supposed it to be caught at the summit of its flight and lodged on the top of a house. Here, then, it rests without motion, but yet not without the power of doing work, and hence not without energy. For we know very well that if we let it fall it will strike the ground with as much velocity, and, therefore, with as much energy, as it had when it was originally projected upwards. Or we may, if we choose, make use of its energy to assist us in driving in a pile, or utilize it in a multitude of ways.

In its lofty position it is, therefore, not without energy, but this is of a quiet nature, and not due in the least to motion. To what, then, is it due? We reply—to the position which the kilogramme occupies at the top of the house. For just as a body in motion is a very different thing (as regards energy) from a body at rest, so is a body at the top of a house a very different thing from a body at the bottom.

To illustrate this, we may suppose that two men of equal activity and strength are fighting together, each having his pile of stones with which he is about to belabour his adversary. One man, however, has secured for himself and his pile an elevated position on the top of a house, while his enemy has to remain content with a position at the bottom. Now, under these circumstances, you can at once tell which of the two will gain the day—evidently the man on the top of the house, and yet not on account of his own superior energy, but rather on account of the energy which he derives from the elevated position of his pile of stones. We thus see that there is a kind of energy derived from position, as well as a kind derived from velocity, and we shall, in future, call the former *energy of position*, and the latter *energy of motion*.

A Head of Water.

35. In order to vary our illustration, let us suppose there are two mills, one with a large pond of water near it and at a high level, while the other has also a pond, but at a lower level than itself. We need hardly ask

which of the two is likely to work—clearly the one with the pond at a low level can derive from it no advantage whatever, while the other may use the high level pond, or head of water, as this is sometimes called, to drive its wheel, and do its work. There is, thus, a great deal of work to be got out of water high up—real substantial work, such as grinding corn or thrashing it, or turning wood or sawing it. On the other hand, there is no work at all to be got from a pond of water that is low down.

A Cross-bow bent. A Watch wound up.

36. In both of the illustrations now given, we have used the force of gravity as that force against which we are to do work, and in virtue of which a stone high up, or a head of water, is in a position of advantage, and has the power of doing work as it falls to a lower level. But there are other forces besides gravity, and, with respect to these, bodies may be in a position of advantage and be able to do work just as truly as the stone, or the head of water, in the case before mentioned.

Let us take, for instance, the force of elasticity, and consider what happens in a cross-bow. When this is bent, the bolt is evidently in a position of advantage with regard to the elastic force of the bow; and when it is discharged, this energy of position of the bolt is converted into energy of motion, just as, when a stone on the top of a house is allowed to fall, its energy of position is converted into that of actual motion.

In like manner a watch wound up is in a position of advantage with respect to the elastic force of the mainspring, and as the wheels of the watch move this is gradually converted into energy of motion.

Advantage of Position.

37. It is, in fact, the fate of all kinds of energy of position to be ultimately converted into energy of motion. The former may be compared to money in a bank, or capital, the latter to money which we are in the act of spending; and just as, when we have money in a bank, we can draw it out whenever we want it, so, in the case of energy of position, we can make use of it whenever we please. To see this more clearly, let us compare together a watermill driven by a head of water, and a windmill driven by the wind. In the one case we may turn on the water whenever it is most convenient for us, but in the other we must wait until the wind happens to blow. The former has all the independence of a rich man; the latter, all the obsequiousness of a poor one. If we pursue the analogy a step further, we shall see that the great capitalist, or the man who has acquired a lofty position, is respected because he has the disposal of a great quantity of energy; and that whether he be a nobleman or a sovereign, or a general in command, he is powerful only from having something which enables him to make use of the services of others. When the man of wealth pays a labouring man to work for him, he is in truth

converting so much of his energy of position into actual energy, just as a miller lets out a portion of his head of water in order to do some work by its means.

Transmutations of Visible Energy.—A Kilogramme shot upwards.

38. We have thus endeavoured to show that there is an energy of repose as well as a living energy, an energy of position as well as of motion; and now let us trace the changes which take place in the energy of a weight, shot vertically upwards, as it continues to rise. It starts with a certain amount of energy of motion, but as it ascends, this is by degrees changed into that of position, until, when it gets to the top of its flight, its energy is entirely due to position.

To take an example, let us suppose that a kilogramme is projected vertically upwards with the velocity of 19·6 metres in one second. According to the formula of Art. 28, it contains 19·6 units of energy due to its actual velocity.

If we examine it at the end of one second, we shall find that it has risen 14·7 metres in height, and has now the velocity of 9·8. This velocity we know (Art. 26) denotes an amount of actual energy equal to 4·9, while the height reached corresponds to an energy of position equal to 14·7. The kilogramme has, therefore, at this moment a total energy of 19·6, of which 14·7 units are due to position, and 4·9 to actual motion.

If we next examine it at the end of another second, we shall find that it has just been brought to rest, so that its energy of motion is *nil;* nevertheless, it has succeeded in raising itself 19·6 metres in height, so that its energy of position is 19·6.

There is, therefore, no disappearance of energy during the rise of the kilogramme, but merely a gradual change from one kind to another. It starts with actual energy, and this is gradually changed into that of position; but if, at any stage of its ascent, we add together the actual energy of the kilogramme, and that due to its position, we shall find that their sum always remains the same.

39. Precisely the reverse takes place when the kilogramme begins its descent. It starts on its downward journey with no energy of motion whatever, but with a certain amount of energy of position; as it falls, its energy of position becomes less, and its actual energy greater, the sum of the two remaining constant throughout, until, when it is about to strike the ground, its energy of position has been entirely changed into that of actual motion, and it now approaches the ground with the velocity, and, therefore, with the energy, which it had when it was originally projected upwards.

The Inclined Plane.

40. We have thus traced the transmutations, as regards energy, of a kilogramme shot vertically upwards, and allowed to fall again to the earth, and we may now

vary our hypothesis by making the kilogramme rise vertically, but descend by means of a smooth inclined plane without friction—imagine in fact, the kilogramme to be shaped like a ball or roller, and the plane to be perfectly smooth. Now, it is well known to all students of dynamics, that in such a case the velocity which the kilogramme has when it has reached the bottom of the plane will be equal to that which it would have had if it had been dropped down vertically through the same height, and thus, by introducing a smooth inclined plane of this kind, you neither gain nor lose anything as regards energy.

In the first place, you do not gain, for think what would happen if the kilogramme, when it reached the bottom of the inclined plane, should have a greater velocity than you gave it originally, when you shot it up. It would evidently be a profitable thing to shoot up the kilogramme vertically, and bring it down by means of the plane, for you would get back more energy than you originally spent upon it, and in every sense you would be a gainer. You might, in fact, by means of appropriate apparatus, convert the arrangement into a perpetual motion machine, and go on accumulating energy without limit—but this is not possible.

On the other hand, the inclined plane, unless it be rough and angular, will not rob you of any of the energy of the kilogramme, but will restore to you the full amount, when once the bottom has been reached. Nor does it matter what be the length or shape of the plane, or whether it be straight, or curved, or spiral, for in all cases, if it only be smooth and of the same vertical height, you will get the same amount of energy by causing the kilogramme to fall from the top to the bottom.

41. But while the energy remains the same, the time of descent will vary according to the length and shape of the plane, for evidently the kilogramme will take a longer time to descend a very sloping plane than a very steep one. In fact, the sloping plane will take longer to generate the requisite velocity than the steep one, but both will have produced the same result as regards energy, when once the kilogramme has arrived at the bottom.

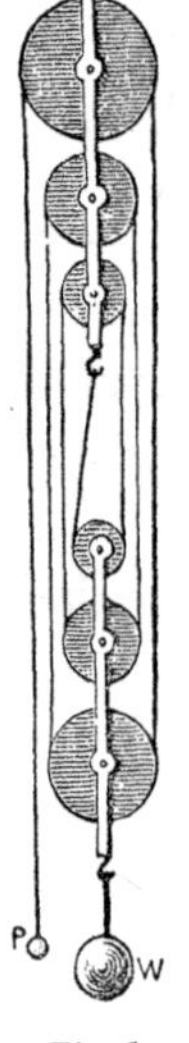

Fig. 1.

Functions of a Machine.

42. Our readers are now beginning to perceive that energy cannot be created, and that by no means can we coax or cozen Dame Nature into giving us back more than we are entitled to get. To impress this fundamental principle still more strongly upon our minds, let us consider in detail one or two mechanical contrivances, and see what they amount to as regards energy.

Let us begin with the second system of pulleys. Here we have a power P attached to the one end of a thread, which passes over all the pulleys, and is ultimately attached, by its other extremity, to a hook in the upper or fixed block. The weight W is, on the other hand, attached to the lower or moveable block, and rises with it. Let us suppose that the pulleys are without weight and the cords without friction, and that W is supported by six cords, as in the figure. Now, when there is equilibrium in this machine, it is well known that W will be equal to six times P; that is to say, a power of one kilogramme will, in such a machine, balance or support a weight of six kilogrammes. If P be increased a single grain more, it will overbalance W, and P will descend, while W will begin to rise. In such a case, after P has descended, say six metres, its weight being, say, one kilogramme, it has lost a quantity of energy of position equal to six units, since it is at a lower level by six metres than it was before. We have, in fact, expended upon our machine six units of energy. Now, what return have we received for this expenditure? Our return is clearly the rise of W, and mechanicians will tell us that in this case W will have risen one metre.

But the weight of W is six kilogrammes, and this having been raised one metre represents an energy of position equal to six. We have thus spent upon our machine, in the fall of P, an amount of energy equal to six units, and obtained in the rise of W an equivalent amount equal to six units also. We have, in truth, neither gained nor lost energy, but simply changed it into a form more convenient for our use.

43. To impress this truth still more strongly, let us take quite a different machine, such as the hydrostatic press. Its mode of action will be perceived from Fig. 2. Here we have two cylinders, a wide and a narrow one, which are connected together at the bottom by means of a strong tube. Each of these cylinders is provided with a water-tight piston, the space beneath being filled with water. It is therefore manifest, since the two cylinders are connected together, and since water is incompressible, that when we push down the one piston the other will be pushed up. Let us suppose that the area of the small piston is one square centimetre,* and that of the large piston one hundred square centimetres, and let us apply a weight of ten kilogrammes to the smaller piston. Now, it is known, from the laws of hydrostatics, that every square centimetre of the larger piston will be pressed upwards with the force of ten kilogrammes, so that the piston will altogether mount with the force of 1000 kilogrammes—that is to say, it will raise a weight of this amount as it ascends.

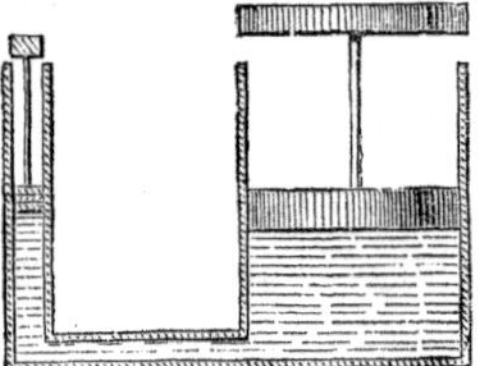
Fig. 2.

Here, then, we have a machine in virtue of which a pressure of ten kilogrammes on the small piston enables the large piston to rise with the force of 1000 kilo-

* That is to say, a square the side of which is one centimetre, or the hundredth part of a metre.

grammes. But it is very easy to see that, while the small piston falls one metre, the large one will only rise one centimetre. For the quantity of water under the pistons being always the same, if this be pushed down one metre in the narrow cylinder, it will only rise one centimetre in the wide one.

Let us now consider what we gain by this machine. The power of ten kilogrammes applied to the smaller piston is made to fall through one metre, and this represents the amount of energy which we have expended upon our machine, while, as a return, we obtain 1000 kilogrammes raised through one single centimetre. Here, then, as in the case of the pulleys, the return of energy is precisely the same as the expenditure, and, provided we ignore friction, we neither gain nor lose anything by the machine. All that we do is to transmute the energy into a more convenient form—what we gain in power we lose in space; but we are willing to sacrifice space or quickness of motion in order to obtain the tremendous pressure or force which we get by means of the hydrostatic press.

Principle of Virtual Velocities.

44. These illustrations will have prepared our readers to perceive the true function of a machine. This was first clearly defined by Galileo, who saw that in any machine, no matter of what kind, if we raise a large weight by means of a small one, it will be found that the small weight, multiplied into the space through which it

is lowered, will exactly equal the large weight, multiplied into that through which it is raised.

This principle, known as that of virtual velocities, enables us to perceive at once our true position. We see that the world of mechanism is not a manufactory, in which energy is created, but rather a mart, into which we may bring energy of one kind and change or barter it for an equivalent of another kind, that suits us better—but if we come with nothing in our hand, with nothing we shall most assuredly return. A machine, in truth, does not create, but only transmutes, and this principle will enable us to tell, without further knowledge of mechanics, what are the conditions of equilibrium of any arrangement.

For instance, let it be required to find those of a lever, of which the one arm is three times as long as the other. Here it is evident that if we overbalance the lever by a single grain, so as to cause the long arm with its power to fall down while the short one with its weight rises up, then the long arm will fall three inches for every inch through which the short arm rises; and hence, to make up for this, a single kilogramme on the long arm will balance three kilogrammes on the short one, or the power will be to the weight as one is to three.

Fig. 3.

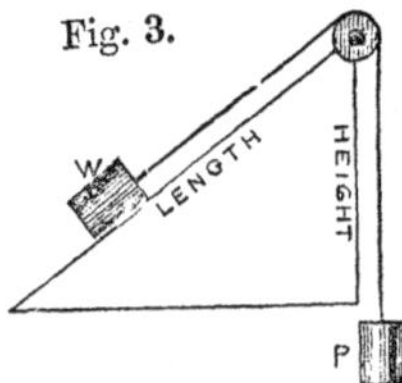

45. Or, again, let us take the inclined plane as represented in Fig. 3.

Here we have a smooth plane and a weight held upon it by means of a power P, as in the figure. Now, if we overbalance P by a single grain, we shall bring the weight W from the bottom to the top of the plane. But when this has taken place, it is evident that P has fallen through a vertical distance equal to the length of the plane, while on the other hand W has only risen through a vertical distance equal to the height. Hence, in order that the principle of virtual velocities shall hold, we must have P multiplied into its fall equal to W multiplied into its rise, that is to say,

$$\text{P} \times \text{Length of plane} = \text{W} \times \text{Height of plane,}$$

$$\text{or } \frac{\text{P}}{\text{W}} = \frac{\text{Height.}}{\text{Length.}}$$

What Friction does.

46. The two examples now given are quite sufficient to enable our readers to see the true function of a machine, and they are now doubtless disposed to acknowledge that no machine will give back more energy than is spent upon it. It is not, however, equally clear that it will not give back less; indeed, it is a well-known fact that it constantly does so. For we have supposed our machine to be without friction—but no machine is without friction—and the consequence is that the available out-come of the machine is more or less diminished by this drawback. Now, unless we are able to see clearly

what part friction really plays, we cannot prove the conservation of energy. We see clearly enough that energy cannot be created, but we are not equally sure that it cannot be destroyed; indeed, we may say we have apparent grounds for believing that it is destroyed—that is our present position. Now, if the theory of the conservation of energy be true—that is to say, if energy is in any sense indestructible—friction will prove itself to be, not the destroyer of energy, but merely the converter of it into some less apparent and perhaps less useful form.

47. We must, therefore, prepare ourselves to study what friction really does, and also to recognize energy in a form remote from that possessed by a body in visible motion, or by a head of water. To friction we may add percussion, as a process by which energy is apparently destroyed; and as we have (Art. 39) considered the case of a kilogramme shot vertically upwards, demonstrating that it will ultimately reach the ground with an energy equal to that with which it was shot upwards, we may pursue the experiment one step further, and ask what becomes of its energy after it has struck the ground and come to rest? We may vary the question by asking what becomes of the energy of the smith's blow after his hammer has struck the anvil, or what of the energy of the cannon ball after it has struck the target, or what of that of the railway train after it has been stopped by friction at the break-wheel? All these

are cases in which percussion or friction appears at first sight to have destroyed visible energy; but before pronouncing upon this seeming destruction, it clearly behoves us to ask if anything else makes its appearance at the moment when the visible energy is apparently destroyed. For, after all, energy may be like the Eastern magicians, of whom we read that they had the power of changing themselves into a variety of forms, but were nevertheless very careful not to disappear altogether.

When Motion is destroyed, Heat appears.

48. Now, in reply to the question we have put, it may be confidently asserted that whenever visible energy is apparently destroyed by percussion or friction, something else makes its appearance, and that something is *heat.* Thus, a piece of lead placed upon an anvil may be greatly heated by successive blows of a blacksmith's hammer. The collision of flint and steel will produce heat, and a rapidly-moving cannon ball, when striking against an iron target, may even be heated to redness. Again, with regard to friction, we know that on a dark night sparks are seen to issue from the break-wheel which is stopping a railway train, and we know, also, that the axles of railway carriages get alarmingly hot, if they are not well supplied with grease.

Finally, the schoolboy will tell us that he is in the habit of rubbing a brass button upon the desk, and applying it to the back of his neighbour's hand, and that when his own hand has been treated in this way, he has found the button unmistakeably hot.

Heat a species of Motion.

49. For a long time this appearance of heat by friction or percussion was regarded as inexplicable, because it was believed that heat was a kind of matter, and it was difficult to understand where all this heat came from. The partisans of the material hypothesis, no doubt, ventured to suggest that in such processes heat might be drawn from the neighbouring bodies, so that the Caloric (which was the name given to the imaginary substance of heat) was squeezed or rubbed out of them, according as the process was percussion or friction. But this was regarded by many as no explanation, even before Sir Humphry Davy, about the end of last century, clearly showed it to be untenable.

50. Davy's experiments consisted in rubbing together two pieces of ice until it was found that both were nearly melted, and he varied the conditions of his experiments in such a manner as to show that the heat produced in this case could not be abstracted from the neighbouring bodies.

51. Let us pause to consider the alternatives to which we are driven by this experiment. If we still choose to regard heat as a substance, since this has not been taken from the surrounding bodies, it must necessarily have been created in the process of friction. But if we choose

to regard heat as a species of motion, we have a simpler alternative, for, inasmuch as the energy of visible motion has disappeared in the process of friction, we may suppose that it has been transformed into a species of molecular motion, which we call heat; and this was the conclusion to which Davy came.

52. About the same time another philosopher was occupied with a similar experiment. Count Rumford was superintending the boring of cannon at the arsenal at Munich, and was forcibly struck with the very great amount of heat caused by this process. The source of this heat appeared to him to be absolutely inexhaustible, and, being unwilling to regard it as the creation of a species of matter, he was led like Davy to attribute it to motion.

53. Assuming, therefore, that heat is a species of motion, the next point is to endeavour to comprehend what kind of motion it is, and in what respects it is different from ordinary visible motion. To do this, let us imagine a railway carriage, full of passengers, to be whirling along at a great speed, its occupants quietly at ease, because, although they are in rapid motion, they are all moving at the same rate and in the same direction. Now, suppose that the train meets with a sudden check;—a disaster is the consequence, and the quiet placidity of the occupants of the carriage is instantly at an end.

Even if we suppose that the carriage is not broken up and its occupants killed, yet they are all in a violent state of excitement; those fronting the engine are driven with force against their opposite neighbours, and are, no doubt, as forcibly repelled, each one taking care of himself in the general scramble. Now, we have only to substitute particles for persons, in order to obtain an idea of what takes place when percussion is converted into heat. We have, or suppose we have, in this act the same violent collision of atoms, the same thrusting forward of A upon B, and the same violence in pushing back on the part of B—the same struggle, confusion, and excitement—the only difference being that particles are heated instead of human beings, or their tempers.

54. We are bound to acknowledge that the proof which we have now given is not a direct one; indeed, we have, in our first chapter, explained the impossibility of our ever seeing these individual particles, or watching their movements; and hence our proof of the assertion that heat consists in such movements cannot possibly be direct. We cannot see that it does so consist, but yet we may feel sure, as reasonable beings, that we are right in our conjecture.

In the argument now given, we have only two alternatives to start with—either heat must consist of a motion of particles, or, when percussion or friction is converted into heat, a peculiar substance called caloric must be created, for if heat be not a species of motion it must necessarily be a species of matter. Now, we have preferred to consider heat as a species of motion to the alternative of supposing the creation of a peculiar kind of matter.

55. Nevertheless, it is desirable to have something to say to an opponent who, rather than acknowledge heat to be a species of motion, will allow the creation of matter. To such an one we would say that innumerable experiments render it certain that a hot body is not sensibly heavier than a cold one, so that if heat be a species of matter it is one that is not subject to the law of gravity. If we burn iron wire in oxygen gas, we are entitled to say that the iron combines with the oxygen, because we know that the product is heavier than the original iron by the very amount which the gas has lost in weight. But there is no such proof that during combustion the iron has combined with a substance called caloric, and the absence of any such proof is enough to entitle us to consider heat to be a species of motion, rather than a species of matter.

Heat a Backward and Forward Motion.

56. We shall now suppose that our readers have assented to our proposition that heat is a species of motion. It is almost unnecessary to add that it must be a species of backward and forward motion; for nothing is more clear than that *a heated substance is not in motion as a whole,* and will not, if put upon a table, push its way from the one end to the other.

Mathematicians express this peculiarity by saying that,

although there is violent internal motion among the particles, yet the centre of gravity of the substance remains at rest; and since, for most purposes, we may suppose a body to act as if concentrated at its centre of gravity, we may say that the body is at rest.

57. Let us here, before proceeding further, borrow an illustration from that branch of physics which treats of sound. Suppose, for instance, that a man is accurately balanced in a scale-pan, and that some water enters his ear; of course he will become heavier in consequence, and if the balance be sufficiently delicate, it will exhibit the difference. But suppose a sound or a noise enters his ear, he may say with truth that something has entered, but yet that something is not matter, nor will he become one whit heavier in consequence of its entrance, and he will remain balanced as before. Now, a man into whose ear sound has entered may be compared to a substance into which heat has entered; we may therefore suppose a heated body to be similar in many respects to a sounding body, and just as the particles of a sounding body move backwards and forwards, so we may suppose that the particles of a heated body do the same.

We shall take another opportunity (Art. 162) to enlarge upon this likeness; but, meanwhile, we shall suppose that our readers perceive the analogy.

Mechanical Equivalent of Heat.

58. We have thus come to the conclusion that when any heavy body, say a kilogramme weight, strikes the ground, the visible energy of the kilogramme is changed into heat; and now, having established the fact of a relationship between these two forms of energy, our next point is to ascertain according to what law the heating effect depends upon the height of fall. Let us, for instance, suppose that a kilogramme of water is allowed to drop from the height of 848 metres, and that we have the means of confining to its own particles and retaining there the heating effect produced. Now, we may suppose that its descent is accomplished in two stages; that, first of all, it falls upon a platform from the height of 424 metres, and gets heated in consequence, and that then the heated mass is allowed to fall other 424 metres. It is clear that the water will now be doubly heated; or, in other words, the heating effect in such a case will be proportional to the height through which the body falls—that is to say, it will be proportional to the actual energy which the body possesses before the blow has changed this into heat. In fact, just as the actual energy represented by a fall from a height is proportional to the height, so is the heating effect, or molecular energy, into which the actual energy is changed proportional to the height also. Having established this point, we now wish to know through

how many metres a kilogramme of water must fall in order to be heated one degree centigrade.

59. For a precise determination of this important point, we are indebted to Dr. Joule, of Manchester, who has, perhaps, done more than any one else to put the science of energy upon a sure foundation. Dr. Joule made numerous experiments, with the view of arriving at the exact relation between mechanical energy and heat; that is to say, of determining the mechanical equivalent of heat. In some of the most important of these he took advantage of the friction of fluids.

60. These experiments were conducted in the following manner. A certain fixed weight was attached to a pulley, as in the figure. The weight had, of course, a tendency

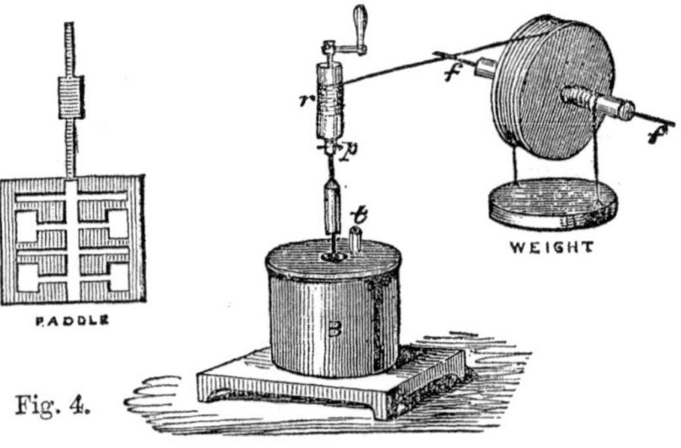

Fig. 4.

to descend, and hence to turn the pulley round. The pulley had its axle supported upon friction wheels, at *f* and *f*, by means of which the friction caused by the movement of the pulley was very much reduced. A string, passing over the circumference of the pulley, was wrapped round *r*, so that, as the weight descended, the pulley moved round, and the string of the pulley caused *r* to rotate very rapidly. Now, the motion of the axis *r* was conducted within the covered box B, where there was attached to *r* a system of paddles, of which a sketch is given in figure; and therefore, as *r* moved, these paddles moved also. There were, altogether, eight sets of these paddles revolving between four stationary vanes. If, therefore, the box were full of liquid, the paddles and the vanes together would churn it about, for these stationary vanes would prevent the liquid being carried along by the paddles in the direction of rotation.

Now, in this experiment, the weight was made to descend through a certain fixed distance, which was accurately measured. As it descended, the paddles were set in motion, and the energy of the descending weight was thus made to churn, and hence to heat some water contained in the box B. When the weight had descended a certain distance, by undoing a small peg *p*, it could be wound up again without moving the paddles in B, and thus the heating effect of several falls of the weight could be accumulated until this became so great as to be capable of being accurately measured by a thermometer. It ought to be mentioned that great care was taken in these experiments, not only to reduce the friction of the axles of the pulley as much as possible, but also to estimate and correct for this friction as accurately as possible; in fact, every precaution was taken to make the experiment successful.

61. Other experiments were made by Joule, in some of which a disc was made to rotate against another disc of cast-iron pressed against it, the whole arrangement being immersed in a cast-iron vessel filled with mercury. From all these experiments, Dr. Joule concluded that the quantity of heat produced by friction, if we can preserve and accurately measure it, will always be found proportional to the quantity of work expended. He expressed this proportion by stating the number of units of work in kilogrammetres necessary to raise by 1° C. the temperature of one kilogramme of water. This was 424, as determined by his last and most complete experiments; and hence we may conclude that if a kilogramme of water be allowed to fall through 424 metres, and if its motion be then suddenly stopped, sufficient heat will be generated to raise the temperature of the water through 1° C., and so on, in the same proportion.

62. Now, if we take the kilogrammetre as our unit of work, and the heat necessary to raise a kilogramme of water 1° C. as our unit of heat, this proportion may be expressed by saying that *one heat unit is equal to* 424 *units of work.*

This number is frequently spoken of as the mechanical equivalent of heat; and in scientific treatises it is denoted by J., the initial of Dr. Joule's name.

63. We have now stated the exact relationship that subsists between mechanical energy and heat, and before proceeding further with proofs of the great law of conservation, we shall endeavour to make our readers acquainted with other varieties of energy, on the ground that it is necessary to penetrate the various disguises that our magician assumes before we can pretend to explain the principles that actuate him in his transformations.

CHAPTER III.

THE FORCES AND ENERGIES OF NATURE: THE LAW OF CONSERVATION.

64. In the last chapter we introduced our readers to two varieties of energy, one of them visible, and the other invisible or molecular; and it will now be our duty to search through the whole field of physical science for other varieties. Here it is well to bear in mind that all energy consists of two kinds, that of *position* and that of *actual motion*, and also that this distinction holds for invisible molecular energy just as truly as it does for that which is visible. Now, energy of position implies a body in a position of advantage with respect to some force, and hence we may with propriety begin our search by investigating the various forces of nature.

Gravitation.

65. The most general, and perhaps the most important, of these forces is *gravitation*, and the law of action of this force may be enunciated as follows:—*Every particle of the universe attracts every other particle with a force*

depending jointly upon the mass of the attracting and of the attracted particle, and varying inversely as the square of distance between the two. A little explanation will make this plain.

Suppose a particle or system of particles of which the mass is unity to be placed at a distance equal to unity from another particle or system of particles of which the mass is also unity—the two will attract each other. Let us agree to consider the mutual attraction between them equal to unity also.

Suppose, now, that we have on the one side two such systems with a mass represented by 2, and on the other side the same system as before, with a mass represented by unity, the distance, meanwhile, remaining unaltered. It is clear the double system will now attract the single system with a twofold force. Let us next suppose the mass of both systems to be doubled, the distance always remaining the same. It is clear that we shall now have a fourfold force, each unit of the one system attracting each unit of the other. In like manner, if the mass of the one system is 2, and that of the other 3, the force will be 6. We may, for instance, call the components of the one system A_1, A_2 and those of the other A_3, A_4, A_5 and we shall have A_1 pulled towards A_3, A_4 and A_5 with a threefold force, and A_2 pulled towards A_3, A_4 and A_5 with a threefold force, making altogether a force equal to 6.

In the next place, let the masses remain unaltered, but let the distance between them be doubled, then the force will be reduced fourfold. Let the distance be tripled, then the force will be reduced ninefold, and so on.

66. Gravitation may be described as a very weak force, capable of acting at a distance, or at least of appearing to do so. It takes the mass of the whole earth to produce the force with which we are so familiar at its surface, and the presence of a large mass of rock or mountain does not produce any appreciable difference in the weight of any substance. It is the gravitation of the earth, lessened of course by distance, which acts upon the moon 240,000 miles away, and the gravitation of the sun influences in like manner the earth and the various other planets of our system.

Elastic Forces.

67. Elastic forces, although in their mode of action very different from gravity, are yet due to visible arrangements of matter; thus, when a cross-bow is bent, there is a visible change produced in the bow, which, as a whole, resists this bending, and tends to resume its previous position. It therefore requires energy to bend a bow, just as truly and visibly as it does to raise a weight above the earth, and elasticity is, therefore, as truly a species of force as gravity is. We shall not here attempt to discuss the various ways in which this force may act, or in which a solid elastic substance will resist

all attempts to deform it; but in all cases it is clearly manifest that work must be spent upon the body, and the force of elasticity must be encountered and overcome throughout a certain space before any sensible deformation can take place.

Force of Cohesion.

68. Let us now leave the forces which animate large masses of matter, and proceed to discuss those which subsist between the smaller particles of which these large masses are composed. And here we must say one word more about molecules and atoms, and the distinction we feel ourselves entitled to draw between these very small bodies, even although we shall never be able to see either the one or the other.

In our first chapter (Art. 7) we supposed the continual sub-division of a grain of sand until we had arrived at the smallest entity retaining all the properties of sand —this we called a *molecule*, and nothing smaller than this is entitled to be called sand. If we continue this sub-division further, the molecule of sand separates itself into its chemical constituents, consisting of silicon on the one side, and oxygen on the other. Thus we arrive at last at the smallest body which can call itself silicon, and the smallest which can call itself oxygen, and we have no reason to suppose that either of these is capable of sub-division into something else, since we regard oxygen and silicon as elementary or simple bodies. Now, these constituents of the silicon molecule are called *atoms*, so that we say the sand molecule is divisible into atoms of silicon and of oxygen. Furthermore, we have strong reason for supposing that such molecules and atoms really exist, but into the arguments for their existence we cannot now enter—it is one of those things that we must ask our readers to take for granted.

69. Let us now take two molecules of sand. These, when near together, have a very strong attraction for each other. It is, in truth, this attraction which renders it difficult to break up a crystalline particle of sand or rock crystal. But it is only exerted when the molecules are near enough together to form a homogeneous crystalline structure, for let the distance between them be somewhat increased, and we find that all attraction entirely vanishes. Thus there is little or no attraction between different particles of sand, even although they are very closely packed together. In like manner, the integrity of a piece of glass is due to the attraction between its molecules; but let these be separated by a flaw, and it will soon be found that this very small increase of distance greatly diminishes the attraction between the particles, and that the structure will now fall to pieces from the slightest cause. Now, these examples are sufficient to show that molecular attraction or *cohesion*, as this is called, is a force which acts very powerfully through a certain small distance, but which vanishes altogether when this distance becomes perceptible. Cohesion is

strongest in solids, while in liquids it is much diminished, and in gases it may be said to vanish altogether. The molecules of gases are, in truth, so far away from one another, as to have little or no mutual attraction, a fact proved by Dr. Joule, whose name was mentioned in the last chapter.

Force of Chemical Affinity.

70. Let us now consider the mutual forces between atoms. These may be characterized as even stronger than the forces between molecules, but as disappearing still more rapidly when the distance is increased. Let us, for instance, take carbon and oxygen—two substances which are ready to combine together to form carbonic acid, whenever they have a suitable opportunity. In this case, each atom of carbon will unite with two of oxygen, and the result will be something quite different from either. Yet under ordinary circumstances carbon, or its representative, coal, will remain unchanged in the presence of oxygen, or of atmospheric air containing oxygen. There will be no tendency to combine together, because although the particles of the oxygen would appear to be in immediate contact with those of the carbon, yet the nearness is not sufficient to permit of chemical affinity acting with advantage. When, however, the nearness becomes sufficient, then chemical affinity begins to operate. We have, in fact, the familiar act of combustion, and, as its consequence, the chemical union of the

carbon or coal with the oxygen of the air, carbonic acid being the result. Here, then, we have a very powerful force acting only at a very small distance, which we name *chemical affinity*, inasmuch as it represents the attraction exerted between atoms of different bodies in contradistinction to cohesion, which denotes the attraction between molecules of the same body.

71. If we regard gravitation as the representative of forces that act or appear to act, at a distance, we may regard cohesion and chemical affinity as the representatives of those forces which, although very powerful, only act or appear to act through a very small interval of distance.

A little reflection will show us how inconvenient it would be if gravitation diminished very rapidly with the distance; for then even supposing that the bond which retains us to the earth were to hold good, that which retains the moon to the earth might vanish entirely, as well as that which retains the earth to the sun, and the consequences would be far from pleasant. Reflection will also show us how inconvenient it would be if chemical affinity existed at all distances; if coal, for instance, were to combine with oxygen without the application of heat, it would greatly alter the value of this fuel to mankind, and would materially check the progress of human industry.

Remarks on Molecular and Atomic Forces.

72. Now, it is important to remember that we must treat cohesion and chemical affinity exactly in the same way as gravity has been treated; and just as we have energy of position with respect to gravity, so may we have as truly a species of energy of position with respect to cohesion and chemical affinity. Let us begin with cohesion.

73. We have hitherto regarded heat as a peculiar motion of the molecules of matter, without any reference to the force which actuates these molecules. But it is a well-known fact that bodies in general expand when heated, so that, in virtue of this expansion, the molecules of a body are driven violently apart against the force of cohesion. Work has in truth been done against this force, just as truly as, when a kilogramme is raised from the earth, work is done against the force of gravity. When a substance is heated, we may, therefore, suppose that the heat has a twofold office to perform, part of it going to increase the actual motions of the molecules, and part of it to separate these molecules from one another against the force of cohesion. Thus, if I swing round horizontally a weight (attached to my hand by an elastic thread of india-rubber), my energy will be spent in two ways—first of all, it will be spent in communicating a velocity to the weight; and, secondly, in stretching the india-rubber string, by means of the

centrifugal tendency of the weight. Work will be done against the elastic force of the string, as well as spent in increasing the motion of the weight.

Now, something of this kind may be taking place when a body is heated, for we may very well suppose heat to consist of a vertical or circular motion, the tendency of which would be to drive the particles asunder against the force of cohesion. Part, therefore, of the energy of heat will be spent in augmenting the motion, and part in driving asunder the particles. We may, however, suppose that, in ordinary cases, the great proportion of the energy of heat goes towards increasing the molecular motion, rather than in doing work against the force of cohesion.

74. In certain cases, however, it is probable that the greater part of the heat applied is spent in doing work against molecular forces, instead of increasing the motions of molecules.

Thus, when a solid melts, or when a liquid is rendered gaseous, a considerable amount of heat is spent in the process, which does not become sensible, that is to say, does not affect the thermometer. Thus, in order to melt a kilogramme of ice, heat is required sufficient to raise a kilogramme of water through 80° C., and yet, when melted, the water is no warmer than the ice. We express this fact by saying that the latent heat of water is 80. Again, if a kilogramme of water at 100° be converted entirely into steam, as much heat is required as would raise the water through 537° C., or 537 kilogrammes of water through one degree; but yet the steam is no hotter than the water, and we express this fact by saying that the latent heat of steam is 537. Now, in both of these instances it is at least extremely probable that a large portion of the heat is spent in doing work against the force of cohesion; and, more especially, when a fluid is converted into a gas, we know that the molecules are in that process separated so far from one another as to lose entirely any trace of mutual force. We may, therefore, conclude that although in most cases the greater portion of the heat applied to a body is spent in increasing its molecular motion, and only a small part in doing work against cohesion, yet when a solid melts, or a liquid vaporizes, a large portion of the heat required is not improbably spent in doing work against molecular forces. But the energy, though spent, is not lost, for when the liquid again freezes, or when the vapour again condenses, this energy is once more transformed into the shape of sensible heat, just as when a stone is dropped from the top of a house, its energy of position is transformed once more into actual energy.

75. A single instance will suffice to give our readers a notion of the strength of molecular forces. If a bar of wrought iron, whose temperature is 10° C. above that of the surrounding medium, be tightly secured at its extremities, it will draw these together with a force of at least one ton for each square inch of section. In some

cases where a building has shown signs of bulging outwards, iron bars have been placed across it, and secured while in a heated state to the walls. On cooling, the iron contracted with great force, and the walls were thereby pulled together.

76. We are next brought to consider atomic forces, or those which lead to chemical union, and now let us see how these are influenced by heat. We have seen that heat causes a separation between the molecules of a body, that is to say, it increases the distance between two contiguous molecules, but we must not suppose that, meanwhile, the molecules themselves are left unaltered.

The tendency of heat to cause separation is not confined to increasing the distance between molecules, but acts also, no doubt, in increasing the distance between parts of the same molecule: in fact, the energy of heat is spent in pulling the constituent atoms asunder against the force of chemical affinity, as well as in pulling the molecules asunder against the force of cohesion, so that, at a very high temperature, it is probable that most chemical compounds would be decomposed, and many are so, even at a very moderate heat.

Thus the attraction between oxygen and silver is so slight that at a comparatively low temperature the oxide of silver is decomposed. In like manner, limestone, or carbonate of lime, is decomposed when subjected to the heat of a lime-kiln, carbonic acid being given off, while quick-lime remains behind. Now, in separating heterogeneous atoms against the powerful force of chemical affinity, work is done as truly as it is in separating molecules from one another against the force of cohesion, or in separating a stone from the earth against the force of gravity.

77. Heat, as we have seen, is very frequently influential in performing this separation, and its energy is spent in so doing; but other energetic agents produce chemical decomposition as well as heat. For instance, certain rays of the sun decompose carbonic acid into carbon and oxygen in the leaves of plants, and their energy is spent in the process; that is to say, it is spent in pulling asunder two such powerfully attracting substances against the affinity they have for one another. And, again, the electric current is able to decompose certain substances, and of course its energy is spent in the process.

Therefore, whenever two powerfully attracting atoms are separated, energy is spent in causing this separation as truly as in separating a stone from the earth, and when once the separation has been accomplished we have a species of energy of position just as truly as we have in a head of water, or in a stone at the top of a house.

78. It is this chemical separation that is meant when we speak of coal as a source of energy. Coal, or carbon, has a great attraction for oxygen, and whenever heat is applied the two bodies unite together. Now oxygen, as it exists in the atmosphere, is the common inheritance of all, and if, in addition to this, some of us possess coal in our cellars, or in pits, we have thus secured a store of

energy of position which we can draw upon with more facility than if it were a head of water, for, although we can draw upon the energy of a head of water whenever we choose, yet we cannot carry it about with us from place to place as we can with coal. We thus perceive that it is not the coal, by itself, that forms the source of energy, but this is due to the fact that we have coal, or carbon, in one place, and oxygen in another, while we have also the means of causing them to unite with each other whenever we wish. If there were no oxygen in the air, coal by itself would be of no value.

Electricity: its Properties.

79. Our readers have now been told about the force of cohesion that exists between molecules of the same body, and also about that of chemical affinity existing between atoms of different bodies. Now, heterogeneity is an essential element of this latter force—there must be a difference of some kind before it can exhibit itself—and under these circumstances its exhibitions are frequently characterized by very extraordinary and interesting phenomena.

We allude to that peculiar exhibition arising out of the forces of heterogenous bodies which we call *electricity,* and, before proceeding further, it may not be out of place to give a short sketch of the mode of action of this very mysterious, but most interesting, agent.

80. The science of electricity is of very ancient origin;

but its beginning was very small. For a couple of thousand years it made little or no progress, and then, during the course of little more than a century, developed into the giant which it now is. The ancient Greeks were aware that amber, when rubbed with silk, had the property of attracting light bodies; and Dr. Gilbert, about three hundred years ago, showed that many other things, such as sulphur, sealing-wax, and glass, have the same property as amber.

In the progress of the science it came to be known that certain substances are able to carry away the peculiar influence produced, while others are unable to do so; the former are called *conductors*, and the latter *non-conductors, or insulators*, of electricity. To make the distinction apparent, let us take a metal rod, having a glass stem attached to it, and rub the glass stem with a piece of silk, care being taken that both silk and glass are warm and dry. We shall find that the glass has now acquired the property of attracting little bits of paper, or elder pith; but only where it has been rubbed, for the peculiar influence acquired by the glass has not been able to spread itself over the surface.

If, however, we take hold of the glass stem, and rub the metal rod, we may, perhaps, produce the same property in the metal, but it will spread over the whole, not confining itself to the part rubbed. Thus we perceive that metal is a conductor, while glass is an insulator, or non-conductor, of electricity.

81. We would next observe that *this influence is of two kinds.* To prove this, let us perform the following experiment. Let us suspend a small pith ball by a very slender silk thread, as in Fig. 5. Next, let us rub a stick of warm, dry glass with a piece of warm silk, and with this excited stick touch the pith ball. The pith ball, after being touched, will be repelled by the excited glass. Let us next excite, in a similar manner, a stick of dry sealing-wax with a piece of warm, dry flannel, and on approaching this stick to the pith ball it will attract it, although the ball, in its present state, is repelled by the excited glass.

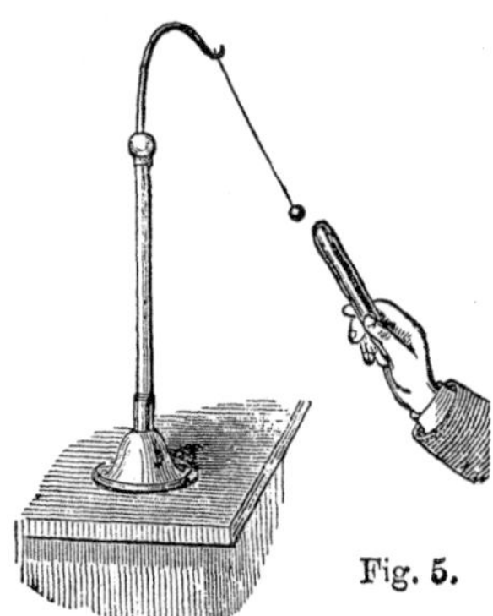

Fig. 5.

Thus a pith ball, touched by excited glass, is repelled by excited glass, but attracted by excited sealing-wax.

In like manner, it might be shown that a pith ball, touched by excited sealing-wax, will be afterwards repelled by excited sealing-wax, but attracted by excited glass.

Now, what the excited glass did to the pith ball, was to communicate to it part of its own influence, after which the ball was repelled by the glass; or, in other words, *bodies charged with similar electricities repel one another.*

Again, since the pith ball, when charged with the electricity from glass, was attracted to the electrified sealing-wax, we conclude that *bodies charged with unlike electricities attract one another.* The electricity from glass is sometimes called *vitreous,* and that from sealing-wax *resinous,* electricity, but more frequently the former is known as *positive,* and the latter as *negative,* electricity—it being understood that these words do not imply the possession of a positive nature by the one influence, or of a negative nature by the other, but are merely terms employed to express the apparent antagonism which exists between the two kinds of electricity.

82. The next point worthy of notice is that *whenever one electricity is produced, just as much is produced of an opposite description.* Thus, in the case of glass excited by silk, we have positive electricity developed upon the glass, while we have also negative electricity developed upon the silk to precisely the same extent. And, again, when sealing-wax is rubbed with flannel, we have negative electricity developed upon the sealing-wax, and just as much positive upon the flannel.

83. These facts have given rise to a theory of electricity, or at least to a method of regarding it, which, if not absolutely correct, seems yet to unite together the various phenomena. According to this hypothesis, a neutral, unexcited body is supposed to contain a store of the two electricities combined together, so that whenever such a body is excited, a separation is produced

between the two. The phenomena which we have described are, therefore, due to this electrical separation, and inasmuch as the two electricities have a great affinity for one another, it requires the expenditure of energy to produce this separation, just as truly as it does to separate a stone from the earth.

84. Now, it is worthy of note that *electrical separation is only produced when heterogeneous bodies are rubbed together.* Thus, if flannel be rubbed upon glass, we have electricity; but if flannel be rubbed upon glass covered with flannel, we have none. In like manner, if silk be rubbed upon sealing-wax covered with silk, or, in fine, if two portions of the same substance be rubbed together, we have no electricity.

On the other hand, a very slight difference of texture is sometimes sufficient to produce electrical separation. Thus, if two pieces of the same silk ribbon be rubbed together lengthwise, we have no electricity; but if they be rubbed across each other, the one is positively, and the other negatively, electrified.

In fact, this element of heterogeneity is an all important one in electrical development, and this leads us to conjecture that *electrical attraction may probably be regarded as peculiarly allied to that force which we call chemical affinity.* At any rate, electricity and chemical affinity are only manifested between bodies that are, in some respects, dissimilar.

85. The following is a list of bodies arranged according

to the electricity which they develop when rubbed together, each substance being positively electrified when rubbed with any substance beneath it in the list.

1. Cat's skin.
2. Flannel.
3. Ivory.
4. Glass.
5. Silk.
6. Wood.
7. Shellac.
8. Resin.
9. Metals.
10. Sulphur.
11. Caoutchouc.
12. Gutta-percha.
13. Gun-cotton.

Thus, if resin be rubbed with cat's skin, or with flannel, the cat's skin or flannel will be positively, and the resin negatively, electrified; while if glass be rubbed with silk, the glass will be positively, and the silk negatively, electrified, and so on.

86. It is not our purpose here to describe at length the *electrical machine*, but we may state that it consists of two parts, one for generating electricity by means of the friction of a rubber against glass, and another consisting of a system of brass tubes, of considerable surface, supported on glass stems, for collecting and retaining the electricity so produced. This latter part of the machine is called its *prime conductor*.

Electric Induction.

87. Let us now suppose that we have set in action a machine of this kind, and accumulated a considerable

quantity of positive electricity in its prime conductor at A. Let us next take two vessels, B and C, made of brass,

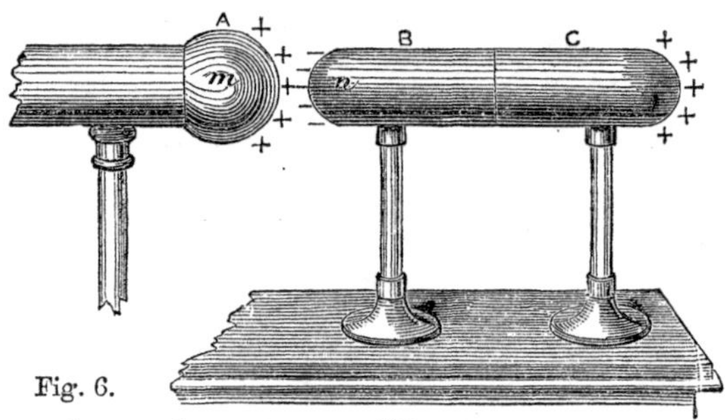

Fig. 6.

supported on glass stems. These two vessels are supposed to be in contact, but at the same time to be capable of being separated from one another at their middle point, where the line is drawn in Fig. 6. Now let us cause B and C to approach A together. At first, B and C are not electrified, that is to say, their two electricities are not separated from each other, but are mixed together; but mark what will happen as they are pushed towards A. The positive electricity of A will decompose the two electricities of B and C, attracting the negative towards itself, and repelling the positive as far away as possible. The disposition of electricities will, therefore, be as in the figure. If we now pull C away from B, we have obtained a quantity of positive electricity on C, by help of the original electricity which was in A; in fact, we have made use of the original stock or electrical capital in A, in order to obtain positive electricity in C, without, however, diminishing the amount of our original stock. Now, this distant action or help, rendered by the original electricity in separating that of B and C, is called electric induction.

88. The experiment may, however, be performed in a somewhat different manner—we may allow B and C to remain together, and gradually push them nearer to A. As B and C approach A, the separation of their electricities will become greater and greater, until, when A and B are only divided by a small thickness of air, the two opposite electricities then accumulated will have sufficient strength to rush together through the air, and unite with each other by means of a spark.

89. The principle of induction may be used with advantage, when it is wished to accumulate a large quantity of electricity.

In this case, an instrument called a *Leyden jar* is very frequently employed. It consists of a glass jar, coated inside and outside with tin foil, as in Fig. 7. A brass rod, having a knob at the end of it, is connected metallically with the inside coating, and is kept in its place by being passed through a cork, which covers the mouth of the jar. We have thus two metallic coatings which are not electrically connected with one another. Now, in order to charge a jar of this kind, let the outside coating be con-

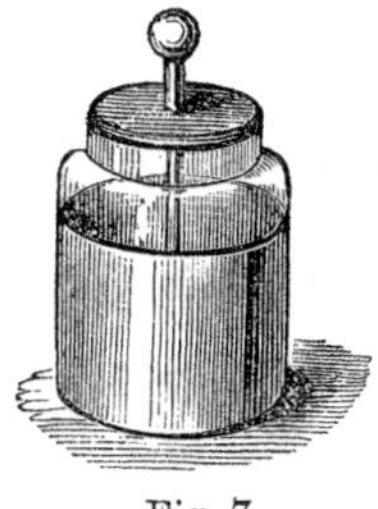

Fig. 7.

nected by a chain with the earth, while at the same time positive electricity from the prime conductor of an electrical machine is communicated to the inside knob.

The positive electricity will accumulate on the inside coating with which the knob is connected. It will then decompose the two electricities of the outside coating, driving the positive electricity to the earth, and there dissipating it, but attracting the negative to itself. There will thus be positive electricity on the inside, and negative on the outside coating. These two electricities may be compared to two hostile armies watching each other, and very anxious to get together, while, however, they are separated from one another by means of an insurmountable obstacle. They will thus remain facing each other, and at their posts, while each side is, meanwhile, being recruited by the same operation as before. We may by this means accumulate a vast quantity of opposite electricities on the two coatings of such a jar, and they will remain there for a long time, especially if the surrounding atmosphere and the glass surface of the jar be quite dry. When, however, electric connection of any kind is made between the two coatings, the electricities rush together and unite with one another in the shape of a spark, while if the human body be the instrument of connecting them a severe shock will be felt.

90. It would thus appear that, when two bodies charged with opposite electricities are brought near each other, the two electricities rush together, forming

a current, and the ultimate result is a spark. Now, this spark implies heat, and is, in truth, nothing else than small particles of intensely heated matter of some kind. We have here, therefore, first of all, the conversion of electrical separation into a current of electricity, and, secondly, the conversion of this current into heat. In this case, however, the current lasts only a very small time; the discharge, as it is called, of a Leyden jar being probably accomplished in $\frac{1}{24000}$th of a second.

The Electric Current.

91. In other cases we have electrical currents which, although not so powerful as that produced by discharging a Leyden jar, yet last longer, and are, in fact, continuous instead of momentary.

We may see a similar difference in the case of visible energy. Thus we might, by means of gunpowder, send up in a moment an enormous mass of water; or we might, by means of a fountain, send up the same mass in the course of time, and in a very much quieter manner. We have the same sort of difference in electrical discharges, and having spoken of the rushing together of two opposite electricities by means of an explosion and a spark, let us now speak of the eminently quiet and effective *voltaic current,* in which we have a continuous coming together of the same two agents.

92. It is not our object here to give a complete description, either historical or scientific, of the voltaic

battery, but rather to give such an account as will enable our readers to understand what the arrangement is, and what sort of effect it produces; and with this object we shall at once proceed to describe the battery of Grove, which is perhaps the most efficacious of all the various arrangements for the purpose of producing an electric current. In this battery we have a number of cells connected together, as in Fig. 8, which shows a battery of three cells. Each cell consists of two vessels, an outer and

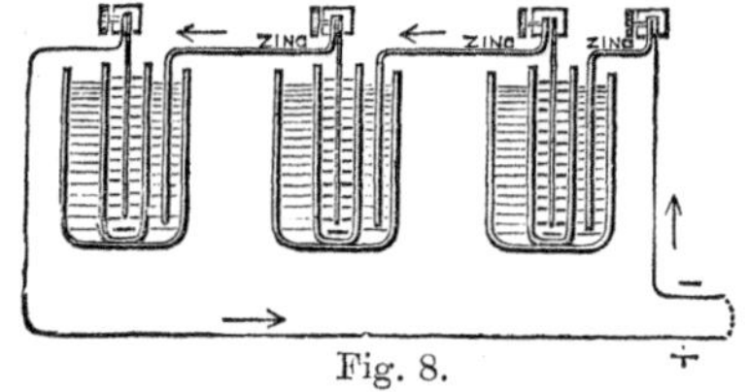

Fig. 8.

an inner one; the outer vessel being made of glass or ordinary stone ware, while the inner one is made of unglazed porcelain, and is therefore porous. The outer vessel is filled with dilute sulphuric acid, and a plate of amalgamated zinc—that is to say, of metallic zinc having its outer surface brightened with mercury,—is immersed in this acid. Again, in the inner or porous vessel we have strong nitric acid, in which a plate of platinum foil is immersed, this being at the same time electrically connected with the zinc plate of the next outer vessel, by means of a clamp, as in the figure. Both metals must be clean where they are pressed together, that is to say, the true metallic surfaces of both must be in contact. Finally, a wire is metallically connected with the platinum of the left-hand cell, and a similar wire with the zinc of the right-hand cell, and these connecting wires ought, except at their extremities, to be covered over with gutta-percha or thread. The loose extremities of these wires are called the *poles* of the battery.

93. Let us now suppose that we have a battery containing a good many cells of this description, and let the whole arrangement be insulated, by being set upon glass supports, or otherwise separated from the earth. If now we test, by appropriate methods, the extremity of the wire connected with the left-hand platinum plate, it will be found to be charged with positive electricity, while the extremity of the other wire will be found charged with negative electricity.

94. In the next place, if we connect these poles of the battery with one another, the two electricities will rush together and unite, or, in other words, there will be an electric current; but it will not be a momentary but a continuous one, and for some time, provided these poles are kept together, a current of electricity will pass through the wires, and indeed through the whole arrangement, including the cells.

The direction of the current will be such that *positive electricity may be supposed to pass from the zinc to the platinum, through the liquid; and back again through the wire, from the platinum at the left hand, to the zinc at the right;* in fact, to go in the direction indicated by the arrow-head.

95. Thus we have two things. In the first place, before

the two terminals, or poles, have been brought together, we have them charged with opposite electricities; and, secondly, when once they have been brought together, we have the production of a continuous current of electricity. Now, this current is an energetic agent, in proof of which we shall proceed to consider the various properties which it has,—the various things which it can do.

Its Magnetic Effects.

96. In the first place, *it can deflect the magnetic needle.* For instance, let a compass needle be swung freely, and let a current of electricity circulate along a wire placed near this needle, and in the direction of its length, then the direction in which the needle points will be immediately altered. This direction will now depend upon that of the current, conveyed by the wire, and the needle will endeavour to place itself at right angles to this wire.

In order to remember the connection between the direction of the current and that of the magnet, imagine your body to form part of the positive current, which may be supposed to enter in at your head, and go out at your feet; also imagine that your face is turned towards the magnet. In this case, the pole of the magnet, which points to the north, will always be deflected by the current towards your right hand. The strength of a current may be measured by the amount of the deflection it produces upon a magnetic needle, and the instrument by which this measurement is made is called a *galvanometer.*

97. In the next place, *the current is able,* not merely to deflect a magnet, but also *to render soft iron magnetic.* Let us take, for instance, the wire connected with the one pole of the battery, and cover it with thread, in order to insulate it, and let us wrap this wire round a cylinder of soft iron, as in Fig. 9. If we now make a communication between the other extremity of the wire, and the other pole of the battery, so as to make the current pass, it will be found that our cylinder of soft iron has become a powerful magnet, and that if an iron keeper be attached to it as in the figure, the keeper will be able to sustain a very great weight.

Fig. 9.

Its Heating Effect.

98. *The electric current has likewise the property of heating a wire through which it passes.* To prove this, let us connect the two poles of a battery by means of a fine platinum wire, when it will be found that the wire will, in a few seconds, become heated to redness. In point of fact, the current will heat a thick wire, but not so much as a thin one, for we may suppose it to rush with great violence through the limited section of the thin wire, producing in its passage great heat.

Its Chemical Effect.

99. Besides its magnetic and heating effects, *the current has also the power of decomposing compound substances*, under certain conditions. Suppose, for instance, that the poles of a battery, instead of being brought together, are plunged into a vessel of water, decomposition will at once begin, and small bubbles of oxygen will rise from the positive pole, while small bubbles of hydrogen will make their appearance at the negative. If the two gases are collected together in a vessel, they may be exploded, and if collected separately, it may be proved by the ordinary tests, that the one is oxygen and the other hydrogen.

Attraction and Repulsion of Currents.

100. We have now described very shortly the magnetic, the heating, and the chemical effects of currents; it remains for us to describe the effects of currents upon one another.

In the first place, suppose that we have two wires which are parallel to one another, and carry currents going in the same direction; and let us further suppose that these wires are capable of moving, then it is found that they will attract one another. If, however, the wires, although parallel, convey currents going in opposite directions, they will then repel one another. A good way of showing this experimentally is to cause two circular currents to float on water. If these currents both go either in the same direction as the hands of a watch, or in the opposite direction, then the two will attract one another; but if the one goes in the one direction, and the other in the other, they will then repel one another.

Attraction and Repulsion of Magnets.

101. Ampère, who discovered this property of currents, has likewise shown us that in very many respects a magnet may be likened to a collection of circular currents all parallel to one another, their direction being such that, if you look towards the north pole of a freely suspended cylindrical magnet facing it, the positive current will descend on the east or left-hand side, and ascend on the west or right-hand side. If we adopt this method of viewing magnets, we can easily account for the attraction between the unlike and the repulsion between the like poles of a magnet, for when unlike poles are placed near each other, the circular currents which face each other are then all going in the same direction, and the two will, therefore, attract one another, but if like poles are placed in this position, the currents that face each other are going in opposite directions, and the poles will, therefore, repel one another.

Induction of Currents.

102. Before closing this short sketch of electrical phenomena, we must allude to the inductive effect of

currents upon each other. Let us suppose (Fig. 10) that we have two circular coils of wire, covered with thread, and placed near each other. Let both the extremities of the right-hand coil be connected with the poles of a battery, so as to make a current of electricity circulate round the coil. On the other hand, let the left-hand coil be connected with a galvanometer, thus enabling us to detect the smallest current of electricity which may pass through this coil. Now, it is found that when we first connect the right-hand coil, so as to pass the battery current through it, a momentary current will pass through the left-hand coil, and will deflect the needle of the

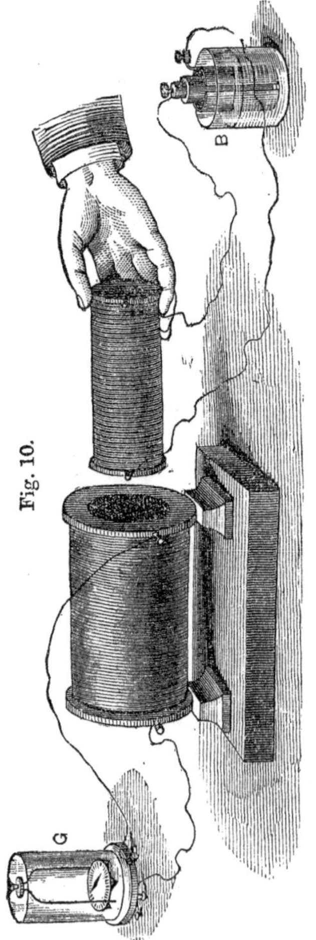

Fig. 10.

galvanometer, but this current will go in an opposite direction to that which circulates round the right-hand coil.

103. Again, as long as the current continues to flow through the right-hand coil there will be no current through the other, but at the moment of breaking the contact between the right-hand coil and the battery there will again be a momentary current in the left-hand coil, but this time in the same direction as that of the right-hand coil, instead of being, as before, in the opposite direction. In other words, when contact is *made* in the right-hand coil, there is a momentary current in the left-hand coil, but in an opposite direction to that in the right, while, when contact is *broken* in the right-hand coil, there is a momentary current in the left-hand coil in the same direction as that in the right.

104. In order to exemplify this induction of currents, it is not even necessary to make and break the current in the right-hand coil, for we may keep it constantly going and arrange so as to make the right-hand coil (always retaining its connection with the battery) alternately approach and recede from the other; when it approaches the other, the effect produced will be the same as when the contact was made in the above experiment—that is to say, we shall have an induced current in an opposite direction to that of the primary, while, when it recedes from the other, we shall have a current in the same direction as that of the primary.

105. Thus we see that whether we keep both coils stationary, and suddenly produce a current in the right-hand coil, or whether, keeping this current constantly going, we suddenly bring it near the other coil, the inductive effect will be precisely the same, for in both cases the left-hand coil is suddenly brought into the presence of a current. And again, it is the same, whether we suddenly break the right-hand current, or suddenly remove it from the left-hand coil, for in both cases this coil is virtually removed from the presence of a current.

List of Energies.

106. We are now in a position to enumerate the various kinds of energy which occur in nature; but, before doing so, we must warn our readers that this enumeration has nothing absolute or complete about it, representing, as it does, not so much the present state of our knowledge as of our want of knowledge, or rather profound ignorance, of the ultimate constitution of matter. It is, in truth, only a convenient classification, and nothing more.

107. To begin, then, with visible energy. We have first of all—

Energy of Visible Motion.

(A.) Visible energy of actual motion—in the planets, in meteors, in the cannon ball, in the storm, in the running stream, and in other instances of bodies in actual visible motion, too numerous to be mentioned.

Visible Energy of Position.

(B.) We have also visible energy of position—in a stone on the top of a cliff, in a head of water, in a rain cloud, in a cross-bow bent, in a clock or watch wound up, and in various other instances.

108. Then we have, besides, several cases in which there is an alternation between (A) and (B).

A pendulum, for instance, when at its lowest point, has only the energy (A), or that of actual motion, in virtue of which it ascends a certain distance against the force of gravity. When, however, it has completed its ascent, its energy is then of the variety (B), being due to position, and not to actual motion; and so on it continues to oscillate, alternately changing the nature of its energy from (A) to (B), and from (B) back again to (A).

109. A vibrating body is another instance of this alternation. Each particle of such a body may be compared to an exceedingly small pendulum oscillating backwards and forwards, only very much quicker than an ordinary pendulum; and just as the ordinary pendulum in passing its point of rest has its energy all of one kind, while in passing its upper point it has it all of another, so when a vibrating particle is passing its point of rest, its energy is all of the variety (A), and when it has reached its extreme displacement, it is all of the variety (B).

Heat Motion.

110. (C.) Coming now to molecular or invisible energy, we have, in the first place, that motion of the molecules of bodies which we term heat. A better term would be *absorbed heat,* to distinguish it from *radiant heat,* which is a very different thing. That peculiar motion which is imparted by heat when absorbed into a body is, therefore, one variety of molecular energy.

Molecular Separation.

(D.) Analagous to this is that effect of heat which represents position rather than actual motion. For part of the energy of absorbed heat is spent in pulling asunder the molecules of the body under the attractive force which binds them together (Art. 73), and thus a store of energy of position is laid up, which disappears again after the body is cooled.

Atomic or Chemical Separation.

111. (E.) The two previous varieties of energy may be viewed as associated with molecules rather than with atoms, and with the force of cohesion rather than with that of chemical affinity. Proceeding now to atomic force, we have a species of energy of position due to the

separation of different atoms under the strong chemical attraction they have for one another. Thus, when we possess coal or carbon and also oxygen in a state of separation from one another, we are in possession of a source of energy which may be called that of chemical separation.

Electrical Separation.

112 (F.) The attraction which heterogeneous atoms possess for one another, sometimes, however, gives rise to a species of energy which manifests itself in a very peculiar form, and appears as electrical separation, which is thus another form of energy of position.

Electricity in Motion.

113 (G.) But we have another species of energy connected with electricity, for we have that due to electricity in motion, or in other words, an electric current which probably represents some form of actual motion.

Radiant Energy.

114 (H.) It is well known that there is no ordinary matter, or at least hardly any, between the sun and the earth, and yet we have a kind of energy

which we may call radiant energy, which proceeds to us from the sun, and proceeds also with a definite, though very great velocity, taking about eight minutes to perform its journey. Now, this radiant energy is known to consist of the vibrations of an elastic medium pervading all space, which is called ether, or the *etherial medium.* Inasmuch, therefore, as it consists of vibrations, it partakes of the character of pendulum motion, that is to say, the energy of any ethereal particle is alternately that of position and that of actual motion.

Law of Conservation.

115. Having thus endeavoured, provisionally at least, to catalogue our various energies, we are in a position to state more definitely what is meant by the conservation of energy. For this purpose, let us take the universe as a whole, or, if this be too large, let us conceive, if possible, a small portion of it to be isolated from the rest, as far as force or energy is concerned, forming a sort of microcosm, to which we may conveniently direct our attention.

This portion, then, neither parts with any of its energy to the universe beyond, nor receives any from it. Such an isolation is, of course, unnatural and impossible, but it is conceivable, and will, at least, tend to concentrate our thoughts. Now, whether we regard the great universe, or this small microcosm, the principle of the conservation of energy asserts that the sum of all the various energies is a constant quantity, that is to say, adopting the language of Algebra—

$(A) + (B) + (C) + (D) + (E) + (F) + (G) + (H) =$ a constant quantity.

116. This does not mean, of course, that (A) is constant in itself, or any other of the left-hand members of this equation, for, in truth, they are always changing about into each other—now, some visible energy being changed into heat or electricity; and, anon, some heat or electricity being changed back again into visible energy—but it only means that the sum of all the energies taken together is constant. We have, in fact, in the left hand, eight variable quantities, and we only assert that their sum is constant, not by any means that they are constant themselves.

117. Now, what evidence have we for this assertion? It may be replied that we have the strongest possible evidence which the nature of the case admits of. The assertion is, in truth, a peculiar one—peculiar in its magnitude, in its universality, in the subtle nature of the agents with which it deals. If true, its truth certainly cannot be proved after the manner in which we prove a proposition in Euclid. Nor does it even admit of a proof so rigid as that of the somewhat analogous principle of the conservation of matter, for in chemistry we may

confine the products of our chemical combination so completely as to prove, beyond a doubt, that no heavy matter passes out of existence that—when coal, for instance, burns in oxygen gas—what we have is merely a change of condition. But we cannot so easily prove that no energy is destroyed in this combination, and that the only result is a change from the energy of chemical separation into that of absorbed heat, for during the process it is impossible to isolate the energy—do what we may, some of it will escape into the room in which we perform the experiment; some of it will, no doubt, escape through the window, while a little will leave the earth altogether, and go out into space. All that we can do in such a case is to estimate, as completely as possible, how much energy has gone away, since we cannot possibly prevent its going. But this is an operation involving great acquaintance with the laws of energy, and very great exactness of observation: in fine, our readers will at once perceive that it is much more difficult to prove the truth of the conservation of energy than that of the conservation of matter.

118. But if it be difficult to prove our principle in the most rigorous manner, we are yet able to give the strongest possible indirect evidence of its truth.

Our readers are no doubt familiar with a method which Euclid frequently adopts in proving his propositions. Starting with the supposition that they are not true, and reasoning upon this hypothesis, he comes to an absurd conclusion—hence he concludes that they are true. Now, we may adopt a method somewhat similar with regard to our principle, only instead of supposing it untrue, let us suppose it true. It may then be shown that, if it be true, under certain test conditions we ought to obtain certain results—for instance, if we increase the pressure, we ought to lower the freezing point of water. Well, we make the experiment, and find that, in point of fact, the freezing point of water is lowered by increasing the pressure, and we have thus derived an argument in favour of the conservation of energy.

119. Or again, if the laws of energy are true, it may be shown that, whenever a substance contracts when heated, it will become colder instead of hotter by compression. Now, we know that ice-cold water, or water just a little above its freezing point, contracts instead of expanding up to 4° C.; and Sir William Thomson has found, by experiment, that water at this temperature is cooled instead of heated by sudden compression. India-rubber is another instance of this relation between these two properties, for if we stretch a string of india-rubber it gets hotter instead of colder, that is to say, its temperature rises by extension, and gets lower by contraction; and again, if we heat the string, we find that it contracts in length instead of expanding like other substances as its temperature increases.

120. Numberless instances occur in which we are

enabled to predict what will happen by assuming the truth of the laws of energy; in other words, these laws are proved to be true in all cases where we can put them to the test of rigorous experiment, and probably we can have no better proof than this of the truth of such a principle. We shall therefore proceed upon the assumption that the conservation of energy holds true in all cases, and give our readers a list of the various transmutations of this subtle agent as it goes backwards and forwards from one abode to another, making, meanwhile, sundry remarks that may tend, we trust, to convince our readers of the truth of our assumption.

CHAPTER IV.

TRANSMUTATIONS OF ENERGY.

Energy of Visible Motion.

121 LET us begin our list of transmutations with the energy of visible motion. This is changed into *energy of position* when a stone is projected upwards above the earth, or, to take a case precisely similar, when a planet or a comet goes from perihelion, or its position nearest the sun, to aphelion, or its position furthest from the sun. We thus see why a heavenly body should move fastest at perihelion, and slowest at aphelion. If, however, a planet were to move round the sun in an orbit exactly circular, its velocity would be the same at all the various points of this orbit, because there would be no change in its distance from the centre of attraction, and therefore no transmutation of energy.

122. We have already (Arts. 108, 109) said that the energy in an oscillating or vibrating body is alternately that of actual motion, and that of position. In this respect, therefore, a pendulum is similar to a comet or heavenly body with an elliptical orbit. Nevertheless the

change of energy is generally more complete in a pendulum or vibrating body than it is in a heavenly body; for in a pendulum, when at its lowest point, the energy is entirely that of actual motion, while at its upper point it is entirely that of position. Now, in a heavenly body we have only a lessening, but not an entire destruction, of the velocity, as the body passes from perihelion to aphelion—that is to say, we have only a partial conversion of the one kind of energy into the other.

123. Let us next consider the change of actual visible energy into *absorbed heat.* This takes place in all cases of friction, percussion, and resistance. In friction, for instance, we have the conversion of work or energy into heat, which is here produced through the rubbing of surfaces against each other; and Davy has shown that two pieces of ice, both colder than the freezing point, may be melted by friction. In percussion, again, we have the energy of the blow converted into heat; while, in the case of a meteor or cannon ball passing through the air with great velocity, we have the loss of energy of the meteor or cannon ball through its contact with the air, and at the same time the production of heat on account of this resistance.

The resistance need not be atmospheric, for we may set the cannon ball to pierce through wooden planks or through sand, and there will equally be a production of heat on account of the resistance offered by the wooden planks or by the sand to the motion of the ball. We may even generalize still further, and assert that whenever the visible momentum of a body is transferred to a larger mass, there is at the same time the conversion of visible energy into heat.

124. A little explanation will be required to make this point clear.

The third law of motion tells us that action and reaction are equal and opposite, so that when two bodies come into collision the forces at work generate equal and opposite quantities of momentum. We shall best see the meaning of this law by a numerical example, bearing in mind that momentum means the product of mass into velocity.

For instance, let us suppose that an inelastic body of mass 10 and velocity 20 strikes directly another inelastic body of mass 15 and velocity 15, the direction of both motions being the same.

Now, it is well known that the united mass will, after impact, be moving with the velocity 17. What, then, has been the influence of the forces developed by collision? The body of greater velocity had before impact a momentum $10 \times 20 = 200$, while its momentum after impact is only $10 \times 17 = 170$; it has therefore suffered a loss of 30 units as regards momentum, or we may consider that a momentum of 30 units has been impressed upon it in an opposite direction to its previous motion.

On the other hand, the body of smaller velocity had before impact a momentum $15 \times 15 = 225$, while after impact it has $15 \times 17 = 255$ units, so that its momentum has been increased by 30 units in its previous direction.

The force of impact has therefore generated 30 units of momentum in two opposite directions, so that, taking account of direction, the momentum of the system is the same before and after impact; for before impact we had a momentum of $10 \times 20 + 15 \times 15 = 425$, while after it we have the united mass 25 moving with the velocity 17, giving the momentum 425 as before.

125. But while the momentum is the same before and after impact, the visible energy of the moving mass is undoubtedly less after impact than before it. To see this we have only to turn to the expression of Art. 28, from which we find that the energy before impact was as follows:—Energy in kilogrammetres $= \frac{m\,v^2}{19 \cdot 6} = \frac{10 \times 20^2 + 15 \times 15^2}{19 \cdot 6} = 376$ nearly; while that after impact $= \frac{25 \times 17^2}{19 \cdot 6} = 368$ nearly.

126. The loss of energy will be still more manifest if we suppose an inelastic body in motion to strike against a similar body at rest. Thus if we have a body of mass 20 and velocity 20 striking against one of equal mass, but at rest, the velocity of the double mass after impact will obviously be only 10; but, as regards energy, that before impact will be $\frac{20 \times 20^2}{19 \cdot 6} = \frac{8000}{19 \cdot 6}$ while that after

impact will be $\frac{40 \times 10^2}{19 \cdot 6} = \frac{4000}{19 \cdot 6}$ or only half the former.

127. Thus there is in all such cases an apparent loss of visible energy, while at the same time there is the production of heat on account of the blow which takes place. If, however, the substances that come together be perfectly elastic (which no substance is), the visible energy after impact will be the same as that before, and in this case there will be no conversion into heat. This, however, is an extreme supposition, and inasmuch as no substance is perfectly elastic, we have in all cases of collision a greater or less conversion of visible motion into heat.

128. We have spoken (Art. 122) about the change of energy in an oscillating or vibrating body, as if it were entirely one of actual energy into energy of position, and the reverse.

But even here, in each oscillation or vibration, there is a greater or less conversion of visible energy into heat. Let us, for instance, take a pendulum, and, in order to make the circumstances as favourable as possible, let it swing on a knife edge, and in vacuo; in this case there will be a slight but constant friction of the knife edge against the plane on which it rests, and though the pendulum may continue to swing for hours, yet it will ultimately come to rest.

And, again, it is impossible to make a vacuum so perfect that there is absolutely no air surrounding the pendulum, so that part of the motion of the pendulum will always be carried off by the residual air of the vacuum in which it swings.

129. Now, something similar happens in that vibratory motion which constitutes sound. Thus, when a bell is in vibration, part of the energy of the vibration is carried off by the surrounding air, and it is in virtue of this that we hear the sound of the bell; but, even if there were no air, the bell would not go on vibrating for ever. For there is in all bodies a greater or less amount of internal viscosity, a property which prevents perfect freedom of vibration, and which ultimately converts vibrations into heat.

A vibrating bell is thus very much in the same position as an oscillating pendulum, for in both part of the energy is given off to the air, and in both there is unavoidable friction—in the one taking the shape of internal viscosity, and in the other that of friction of the knife edge against the plane on which it rests.

130. In both these cases, too, that portion of the energy which goes into the air takes ultimately the shape of heat. The oscillating pendulum communicates a motion to the air, and this motion ultimately heats the air. The vibrating bell, or musical instrument, in like manner communicates part of its energy to the air. This communicated energy first of all moves through the air with the well-known velocity of sound, but during its progress it, too, no doubt becomes partly converted into heat. Ultimately, it is transmitted by the air to other bodies,

and by means of their internal viscosity is sooner or later converted into heat. Thus we see that heat is the form of energy, into which all visible terrestrial motion, whether it be rectilinear, or oscillatory, or vibratory, is ultimately changed.

131. In the case of a body in visible rectilinear motion on the earth's surface, this change takes place very soon—if the motion be rotatory, such as that of a heavy revolving top, it may, perhaps, continue longer before it is ultimately stopped, by means of the surrounding air, and by friction of the pivot; if it be oscillatory, as in the pendulum, or vibratory, as in a musical instrument, we have seen that the air and internal friction are at work, in one shape or another, to carry it off, and will ultimately succeed in converting it into heat.

132. But, it may be said, why consider a body moving on the earth's surface? why not consider the motion of the earth itself? Will this also ultimately take the shape of heat?

No doubt it is more difficult to trace the conversion in such a case, inasmuch as it is not proceeding at a sensible rate before our eyes. In other words, the very conditions that make the earth habitable, and a fit abode for intelligent beings like ourselves, are those which unfit us to perceive this conversion of energy in the case of the earth. Yet we are not without indications that it is actually taking place. For the purpose of exhibiting these, we may divide the earth's motion into two—a motion of rotation, and one of revolution.

133. Now, with regard to the earth's rotation, the conversion of the visible energy of this motion into heat is already well recognized. To understand this we have only to study the nature of the moon's action upon the fluid portions of our globe. In the following diagram (Fig. 11) we have an exaggerated representation of this, by which we see that the spherical earth is converted

Fig. 11.

into an elongated oval, of which one extremity always points to the moon. The solid body of the earth itself revolves as usual, but, nevertheless, this fluid protuberance remains always pointing towards the moon, as we see in the figure, and hence the earth rubs against the protuberance as it revolves. The friction produced by this action tends evidently to lessen the rotatory energy of the earth—in other words, it acts like a break—and we have, just as by a break-wheel, the conversion of visible energy into heat. This was first recognized by Mayer and J. Thomson.

134. But while there can be no doubt about the fact of such a conversion going on, the only question is regarding

its rate of progress, and the time required before it can cause a perceptible impression upon the rotative energy of the earth.

Now, it is believed by astronomers that they have detected evidence of such a change, for our knowledge of the motions of the sun and moon has become so exact, that not only can we carry forward our calculations so as to predict an eclipse, but also carry them backwards, and thus fix the dates and even the very details of the ancient historical eclipses.

If, however, between those times and the present, the earth has lost a little rotative energy on account of this peculiar action of the moon, then it is evident that the calculated circumstances of the ancient total eclipse will not quite agree with those actually recorded; and by a comparison of this nature it is believed that we have detected a very slight falling off in the rotative energy of our earth. If we carry out the argument, we shall be driven to the conclusion that the rotative energy of our globe will, on account of the moon's action, always get less and less, until things are brought into such a state that the rotation comes to be performed in the same time as the revolution of the moon, so that then the same portion of the terrestrial surface being always presented to the moon, it is evident that there will be no effort made by the solid substance of the earth, to glide from under the fluid protuberance, and there will in consequence be no friction, and no further loss of energy.

135. If the fate of the earth be ultimately to turn the same face always to the moon, we have abundant evidence that this very fate has long since overtaken the moon herself. Indeed, the much stronger effect of our earth upon the moon has produced this result, probably, even in those remote periods when the moon was chiefly fluid; and it is a fact well known, not merely to astronomers, but to all of us, that the moon nowadays turns always the same face to the earth.* No doubt this fate has long since overtaken the satellites of Jupiter, Saturn, and the other large planets; and there are independent indications that, at least in the case of Jupiter, the satellites turn always the same face to their primary.

136. To come now to the energy of revolution of the earth, in her orbit round the sun, we cannot help believing that there is a material medium of some kind between the sun and the earth; indeed, the undulatory theory of light requires this belief. But if we believe in such a medium, it is difficult to imagine that its presence will not ultimately diminish the motion of revolution of the earth in her orbit; indeed, there is a strong scientific probability, if not an absolute certainty, that such will be the case. There is even some reason to think that the energy of a comet of small period, called Encke's comet, is gradually being stopped from this cause; in fine, there can be hardly any doubt that the cause is really in operation,

* This explanation was first given by Professors Thomson and Tait in their Natural Philosophy, and by Dr. Frankland in a lecture at the Royal Institution of London.

and will ultimately affect the motions of the planets and other heavenly bodies, even although its rate of action may be so slow that we are not able to detect it.

We may perhaps generalize by saying, that wherever in the universe there is a differential motion, that is to say, a motion of one part of it towards or from another, then, in virtue of the subtle medium, or cement, that binds the various parts of the universe together, this motion is not unattended by something like friction, in virtue of which the differential motion will ultimately disappear, while the loss of energy caused by its disappearance will assume the form of heat.

137. There are, indeed, obscure intimations that a conversion of this kind is not improbably taking place in the solar system; for, in the sun himself, we have the matter near the equator, by virtue of the rotation of our luminary, carried alternately towards and from the various planets. Now, it would seem that the sun-spots, or atmospheric disturbances of the sun, affect particularly his equatorial regions, and have likewise a tendency to attain their maximum size in that position, which is as far away as possible from the influential planets, such as Mercury or Venus;* so that if Venus, for instance, were between the earth and the sun, there would be few sun-spots in the middle of the sun's disc, because that would be the part of the sun nearest Venus.

* *See* De La Rue, Stewart, and Loewy's researches on Solar Physics.

But if the planets influence sun-spots, the action is no doubt reciprocal, and we have much reason to believe that sun-spots influence, not only the magnetism, but also the meteorology of our earth, so that there are most displays of the Aurora Borealis, as well as most cyclones, in those years when there are most sun-spots. * Is it not then possible that, in these strange, mysterious phenomena, we see traces of the machinery by means of which the differential motion of the solar system is gradually being changed into heat?

138. We have thus seen that visible energy of actual motion is not unfrequently changed into visible energy of position, and that it is also very often transformed into absorbed heat. We have now to state that it may likewise be transformed into *electrical separation.* Thus, when an ordinary electrical machine is in action, considerable labour is spent in turning the handle; it is, in truth, harder to turn than if no electricity were being produced—in other words, part of the energy which is spent upon the machine goes to the production of electrical separation. There are other ways of generating electricity besides the frictional method. If, for instance, we bring an insulated conducting plate near the prime conductor of the electrical machine, yet not near enough to cause a spark to pass, and if we then touch the insulated plate, we shall find it, after contact, to be charged with an electricity the oppo-

* *See* the Magnetic Researches of Sir E. Sabine, also C. Meldrum on the Periodicity of Cyclones.

site of that in the machine; we may then remove it and make use of this electricity.

It requires a little thought to see what labour we have spent in this process. We must bear in mind that, by touching the plate, we have carried off the electricity of the same name as that of the machine, so that, after touching the insulated plate it is more strongly attracted to the conductor than it was before. When we begin to remove it, therefore, it will cost us an effort to do so, and the mechanical energy which we spend in removing it will account for the energy of electrical separation which we then obtain.

139. We may thus make use of a small nucleus of electricity, to assist us in procuring an unlimited supply, for in the above process the electricity of the prime conductor remains unaltered, and we may repeat the operation as often as we like, and gather together a very large quantity of electricity, without finally altering the electricity of the prime conductor, but not, however, without the expenditure of an equivalent amount of energy, in the shape of actual work.

140. While, as we have seen, there is a tendency in all motion to be changed into heat, there is one instance where it is, in the first place at least, changed into *a current of electricity.* We allude to the case where a conducting substance moves in the presence of an electric current, or of a magnet.

In Art. 104 we found that if one coil connected with a

battery were quickly moved into the presence of another coil connected with a galvanometer, an induced current would be generated in the latter coil, and would affect the galvanometer, its direction being the reverse of that passing in the other. Now, an electric current implies energy, and we may therefore conclude that some other form of energy must be spent, or disappear, in order to produce the current which is generated in the coil attached to the galvanometer.

Again, we learn from Art. 100 that two currents going in opposite directions repel one another. The current generated in the coil attached to the galvanometer or secondary current will, therefore, repel the primary current, which is moving towards it; this repulsion will either cause a stoppage of motion, or render necessary the expenditure of energy, in order to keep up the motion of this moving coil. We thus find that two phenomena occur simultaneously. In the first place, there is the production of energy in the secondary coil, in the shape of a current opposite in direction to that of the primary coil; in the next case, owing to the repulsion between this induced current and the primary current, there is a stoppage or disappearance of the energy of actual motion of the moving coil. We have, in fact, the creation of one species of energy, and at the same time the disappearance of another, and thus we see that the law of conservation is by no means broken.

141. We see also the necessary connection between the

two electrical laws described in Arts. 100 and 104. Indeed, had these laws been other than what they are, the principle of conservation of energy would have been broken.

For instance, had the induced current in the case now mentioned been in the same direction as that of the primary, the two currents would have attracted each other, and thus there would have been the creation of a secondary current, implying energy, in the coil attached to the galvanometer, along with an increase of the visible energy of motion of the primary current—that is to say, instead of the creation of one kind of energy, accompanied with the disappearance of another, we should have had the simultaneous creation of both; and thus the law of conservation of energy would have been broken.

We thus see that the principle of conservation enables us to deduce the one electrical law from the other, and this is one of the many instances which strengthen our belief in the truth of the great principle for which we are contending.

142. Let us next consider what will take place if we cause the primary current to move from the secondary coil instead of towards it.

In this case we know, from Art. 104, that the induced current will be in the same direction as the primary, while we are told by Art. 100 that the two currents will now attract each other. The tendency of this attraction will be to stop the motion of the primary current from the secondary one, or, in other words, there will be a disappearance of the energy of visible motion, while at the same time there is the production of a current. In both cases, therefore, one form of energy disappears while another takes its place, and in both there will be a very perceptible resistance experienced in moving the primary coil, whether towards the secondary or from it. Work will, in fact, have to be spent in both operations, and the outcome of this work or energy will be the production of a current in the first place, and of heat in the second; for we learn from Art. 98 that when a current passes along a wire its energy is generally spent in heating the wire.

We have thus two phenomena occurring together. In the first place, in moving a current of electricity to and from a coil of wire, or any other conductor, or (which is the same thing, since action and reaction are equal and opposite) in moving a coil of wire or any other conductor to and from a current of electricity, a sense of resistance will be experienced, and energy will have to be spent upon the process; in the second place, an electrical current will be generated in the conductor, and the conductor will be heated in consequence.

143. The result will be rendered very prominent if we cause a metallic top, in rapid rotation, to spin near two iron poles, which, by means of the battery, we can suddenly convert into the poles of a powerful electro-

magnet. When this change is made, and the poles become magnetic, the motion of the top is very speedily brought to rest, just as if it had to encounter a species of invisible friction. This curious result can easily be explained. We have seen from Art. 101 that a magnet resembles an assemblage of electric currents, and in the metallic top we have a conductor alternately approaching these currents and receding from them ; and hence, according to what has been said, we shall have a series of secondary currents produced in the conducting top which will stop its motion, and which will ultimately take the shape of heat. In other words, the visible energy of the top will be changed into heat just as truly as if it were stopped by ordinary friction.

144. The electricity induced in a metallic conductor, moved in the presence of a powerful magnet, has received the name of Magneto-Electricity; and Dr. Joule has made use of it as a convenient means of enabling him to determine the mechanical equivalent of heat, for it is into heat that the energy of motion of the conductor is ultimately transformed. But, besides all this, these currents form, perhaps, the very best means of obtaining electricity ; and recently very powerful machines have been constructed by Wild and others with this view.

145. These machines, when large, are worked by a steam-engine, and their mode of operation is as follows:— The nucleus of the machine is a system of powerful permanent steel magnets, and a conducting coil is made

to revolve rapidly in presence of these magnets. The current produced by this moving coil is then used in order to produce an extremely powerful electro-magnet, and finally a coil is made to move with great rapidity in presence of this powerful electro-magnet, thus causing induced currents of vast strength. So powerful are these currents, that when used to produce the electric light, small print may be read on a dark night at the distance of two miles from the scene of operation!

It thus appears that in this machine a double use is made of magneto-electricity. Starting with a nucleus of permanent magnetism, the magneto-electric currents are used, in the first instance, to form a powerful electro-magnet much stronger than the first, and this powerful electro-magnet is again made use of in the same way as the first, in order to give, by means of magneto-electricity, an induced current of very great strength.

146. There is, moreover, a very great likeness between a magneto-electric machine like that of Wild's for generating electric currents, and the one which generates statical electricity by means of the method already described Art. 139. In both cases advantage is taken of a nucleus, for in the magneto-electric machine we have the molecular currents of a set of permanent magnets, which are made the means of generating enormous electric currents without any permanent alteration to themselves, yet not without the expenditure of work.

Again, in an induction machine for generating statical

electricity, we have an electric nucleus, such as we have supposed to reside in the prime conductor of a machine; and advantage may be taken, as we have seen, of this nucleus in order to generate a vast quantity of statical electricity, without any permanent alteration of the nucleus, but not without the expenditure of work.

147. We have now seen under what conditions the visible energy of actual motion may be changed—1stly, into energy of position; 2ndly, into the two energies which embrace absorbed heat; 3rdly, into electrical separation; and finally into electricity in motion. As far as we know, visible energy cannot directly be transformed into chemical separation, or into radiant energy.

Visible Energy of Position.

148. Having thus exhausted the transmutations of the energy of visible motion, we next turn to that of position, and find that it is transmuted into motion, but not immediately into any other form of energy; we may, therefore, dismiss this variety at once from our consideration.

Absorbed Heat.

149. Coming now to these two forms of energy which embrace *absorbed heat*, we find that this may be converted into (A) or *actual visible energy* in the case of the steam-engine, the air-engine, and all varieties of heat engines. In the steam-engine, for instance, part of the

heat which passes through it disappears as heat, utterly and absolutely, to reappear as mechanical effect. There is, however, one condition which must be rigidly fulfilled, whenever heat is changed into mechanical effect—there must be a difference of temperature, and *heat will only be changed into work, while it passes from a body of high temperature to one of low.*

Carnot, the celebrated French physicist, has ingeniously likened the mechanical power of heat to that of water; for just as you can get no work out of heat unless there be a flow of heat from a higher temperature level to a lower, so neither can you get work out of water unless it be falling from a higher level to a lower.

150. If we reflect that heat is essentially distributive in its nature, we shall soon perceive the reason for this peculiar law; for, in virtue of its nature, heat is always rushing from a body of high temperature to one of low, and if left to itself it would distribute itself equally amongst all bodies, so that they would ultimately become of the same temperature. Now, if we are to coax work out of heat, we must humour its nature, for it may be compared to a pack of schoolboys, who are always ready to run with sufficient violence out of the schoolroom into the open fields, but who have frequently to be dragged back with a very considerable expenditure of energy. So heat will not allow itself to be confined, but will resist any attempt to accumulate it into a limited space. Work cannot, therefore, be gained by

such an operation, but must, on the contrary, be spent upon the process.

151. Let us now for a moment consider the case of an enclosure in which everything is of the same temperature. Here we have a dull dead level of heat, out of which it will be impossible to obtain the faintest semblance of work. The temperature may even be high, and there may be immense stores of heat energy in the enclosure, but not a trace of this is available in the shape of work. Taking up Carnot's comparison, the water has already fallen to the same level, and lies there without any power of doing useful work—dead, in a sense, as far as visible energy is concerned.

152. We thus perceive that, firstly, we can get work out of heat when it passes from a higher to a lower temperature, but that, secondly, we must spend work upon it in order to make it pass from a lower temperature to a higher one ; and that, thirdly and finally, nothing in the shape of work can be got out of heat which is all at the same temperature level.

What we have now said enables us to realize the conditions under which all heat engines work. The essential point about such engines is, not the possession of a cylinder, or piston, or fly wheels, or valves, but the possession of two chambers, one of high and the other of low temperature, while it performs work in the process of carrying heat from the chamber of high to that of low temperature.

Let us take, for example, the low-pressure engine. Here we have the boiler or chamber of high, and the condenser or chamber of low, temperature, and the engine works while heat is being carried from the boiler to the condenser—never while it is being carried from the condenser to the boiler.

In like manner in the locomotive we have the steam generated at a high temperature and pressure, and cooled by injection into the atmosphere.

153. But, leaving formal engines, let us take an ordinary fire, which plays in truth the part of an engine, as far as energy is concerned. We have here the cold air streaming in over the floor of the room, and rushing into the fire, to be there united with carbon, while the rarefied product is carried up the chimney. Dismissing from our thoughts at present the process of combustion, except as a means of supplying heat, we see that there is a continual in-draught of cold air, which is heated by the fire, and then sent to mingle with the air above. Heat is, in fact, distributed by this means, or carried from a body of high temperature, *i.e.* the fire, to a body of low temperature, *i.e.* the outer air, and in this process of distribution mechanical effect is obtained in the up-rush of air through the chimney with considerable velocity.

154. Our own earth is another instance of such an engine, having the equatorial regions as its boiler, and the polar regions as its condensers; for, at the equator, the air is heated by the direct rays

of the sun, and we have there an ascending current of air, up a chimney as it were, the place of which is supplied by an in-draught of colder air along the ground or floor of the world, from the poles on both sides. Thus the heated air makes its way from the equator to the poles in the upper regions of the atmosphere, while the cold air makes its way from the poles to the equator along the lower regions. Very often, too, aqueous vapour as well as air is carried up by means of the sun's heat to the upper and colder atmospheric regions, and there deposited in the shape of rain, or hail, or snow, which ultimately finds its way back again to the earth, often displaying in its passage immense mechanical energy. Indeed, the mariner who hoists his sail, and the miller who grinds his corn (whether he use the force of the wind or that of running water), are both dependent upon this great earth-engine, which is constantly at work producing mechanical effect, but always in the act of carrying heat from its hotter to its colder regions.

155. Now, if it be essential to an engine to have two chambers, one hot and one cold, it is equally important that there should be a considerable temperature difference between the two.

If Nature insists upon a difference before she will give us work, we shall not be able to pacify her, or to meet her requirements by making this difference as small as possible. And hence, *cæteris paribus*, we shall obtain a greater proportion of work out of a certain amount of heat passing through our engine when the temperature difference between its boiler and condenser is as great as possible. In a steam-engine this difference cannot be very great, because if the water of the boiler were at a very high temperature the pressure of its steam would become dangerous; but in an air-engine, or engine that heats and cools air, the temperature difference may be much larger. There are, however, practical inconveniences in engines for which the temperature of the boiler is very high, and it is possible that these may prove so formidable as to turn the scale against such engines, although in theory they ought to be very economical.

156. The principles now stated have been employed by Professor J. Thomson, in his suggestion that the application of pressure would be found to lower the freezing point of water; and the truth of this suggestion was afterwards proved by Professor Sir W. Thomson. The following was the reasoning employed by the former:—

Suppose that we have a chamber kept constantly at the temperature 0° C., or the melting point of ice, and that we have a cylinder, of which the sectional area is one square metre, filled one metre in height with water, that is to say, containing one cubic metre of water. Suppose, next, that a well-fitting piston is placed above the surface of the water in this cylinder, and that a considerable weight is placed upon the piston. Let us now take the cylinder, water and all, and carry it into another room, of which the temperature is just a trifle lower. In course of time the water will freeze, and, as it expands in freezing, it will push up the piston and weight about $\frac{9}{100}$ths of a metre; and we may suppose that the piston is kept fastened in this position by means of a peg. Now carry back the machine into the first room, and in the course of time the ice will be melted, and we shall have water once more in the cylinder, but there will now be a void space of $\frac{9}{100}$ths of a metre between the piston and the surface. We have thus acquired a certain amount of energy of position, and we have only to pull out the peg, and allow the piston with its weight to fall down through the vacant space, in order to utilize this energy, after which the arrangement is ready to start afresh. Again, if the weight be very great, the energy thus gained will be very great; in fact, the energy will vary with the weight. In fine, the arrangement now described is a veritable heat engine, of which the chamber at 0° C. corresponds to the boiler, and the other chamber a trifle lower in temperature to the condenser, while the amount of work we get out of the engine—or, in other words, its efficiency—will depend upon the weight which is raised through the space of $\frac{9}{100}$ths of a metre, so that, by increasing this weight without limit, we may increase the efficiency of our engine without limit. It would thus at first sight appear that by this device of having two chambers, one at 0° C., and the other a trifle lower, we can get any amount of work out of our water engine; and that, consequently, we have managed to overcome

Nature. But here Thomson's law come into operation, showing that we cannot overcome Nature by any such device, but that if we have a large weight upon our piston, we must have a proportionally large difference of temperature between our two chambers—that is to say, the freezing point of water, under great pressure, will be lower in temperature than its freezing point, if the pressure upon it be only small.

Before leaving this subject we must call upon our readers to realize what takes place in all heat engines. It is not merely that heat produces mechanical effect, but that *a given quantity of heat absolutely passes out of existence as heat in producing its equivalent of work.* If, therefore, we could measure the mere heat produced in an engine by the burning of a ton of coals, we should find it to be less when the engine was doing work than when it was at rest.

In like manner, when a gas expands suddenly its temperature falls, because a certain amount of its heat passes out of existence in the act of producing mechanical effect.

157. We have thus endeavoured to show under what conditions absorbed heat may be converted into mechanical effect. This absorbed heat embraces (Art. 110) two varieties of energy, one of these being molecular motion, and the other molecular energy of position.

Let us now, therefore, endeavour to ascertain under what circumstances the one of these varieties may be changed into the other. It is well known that it takes a good deal of heat to convert a kilogramme of ice into water, and that when the ice is melted the temperature of the water is not perceptibly higher than that of the ice. It is equally well known that it takes a great deal of heat to convert a kilogramme of boiling water into steam, and that when the transformation is accomplished, the steam produced is not perceptibly hotter than the boiling water. In such cases the heat is said to become latent.

Now, in both these cases, but more obviously in the last, we may suppose that the heat has not had its usual office to perform, but that, instead of increasing the motion of the molecules of water, it has spent its energy in tearing them asunder from each other, against the force of cohesion which binds them together.

Indeed, we know as a matter of fact that the force of cohesion which is perceptible in boiling water is apparently absent from steam, or the vapour of water, because its molecules are too remote from one another to allow of this force being appreciable. We may, therefore, suppose that a large part, at least, of the heat necessary to convert boiling water into steam is spent in doing work against molecular forces.

When the steam is once more condensed into hot water, the heat thus spent reassumes the form of molecular motion, and the consequence is that we require to take away somehow all the latent heat of a kilogramme of steam before we can convert it into boiling water. In fact, if it is difficult and tedious to convert water into steam, it is difficult and tedious to convert steam into water.

158. Besides the case now mentioned, there are other instances in which, no doubt, molecular separation becomes gradually changed into heat motion. Thus, when a piece of glass has been suddenly cooled, its particles have not had time to acquire their proper position, and the consequence is that the whole structure is thrown into a state of constraint. In the course of time such bodies tend to assume a more stable state, and their particles gradually come closer together.

It is owing to this cause that the bulb of a thermometer recently blown gradually contracts, and it is no doubt owing to the same cause that a Prince Rupert's drop, formed by dropping melted glass into water, when broken, falls into powder with a kind of explosion. It seems probable that in all such cases these changes are attended with heat, and that they denote the conversion of the energy of molecular separation into that of molecular motion.

159. Having thus examined the transmutations of (C) into (D), and of (D) back again into (C), let us now proceed with our list, and see under what circumstances absorbed heat is changed into *chemical separation.*

It is well known that when certain bodies are heated, they are decomposed; for instance, if limestone or carbonate of lime be heated, it is decomposed, the carbonic acid being given out in the shape of gas, while quicklime remains behind. Now, heat is consumed in this process, that is to say, a certain amount of heat energy absolutely passes out of existence *as heat* and is changed into the energy of chemical separation. Again, if the lime so obtained be exposed, under certain circumstances, to an atmosphere of carbonic acid, it will gradually become changed into carbonate of lime; and in this change (which is a gradual one) we may feel assured that the energy of chemical separation is once more converted into the energy of heat, although we may not perceive any increment of temperature, on account of the slow nature of the process.

At very high temperatures it is possible that most compounds are decomposed, and the temperature at which this takes place, for any compound, has been termed its *temperature of disassociation.*

160. Heat energy is changed into *electrical separation* when tourmalines and certain other crystals are heated.

Let us take, for instance, a crystal of tourmaline and raise its temperature, and we shall find one end positively, and the other negatively, electrified. Again, let us take the same crystal, and suddenly cool it, and we shall find an electrification of the opposite kind to the former, so that the end of the axis, which was then positive, will now be negative. Now, this separation of the electricities denotes energy; and we have, therefore, in such crystals a case where the energy of heat has been changed into that of electrical separation. In other words, a certain amount of heat has passed out of existence *as heat*, while in its place a certain amount of electrical separation has been obtained.

161. Let us next see under what circumstances heat is changed into *electricity in motion*. This transmutation takes place in thermo-electricity.

Suppose, for instance, that we have a bar of copper or antimony, say copper, soldered to a bar of bismuth, as in Fig. 12. Let us now heat one of the junctions, while the other remains cool. It will be found that a current of positive electricity circulates round the bar, in the direction of the arrow-head, going from the bismuth to the copper across the heated junction, the existence of which may be detected by means of a compass needle, as we see in the figure.

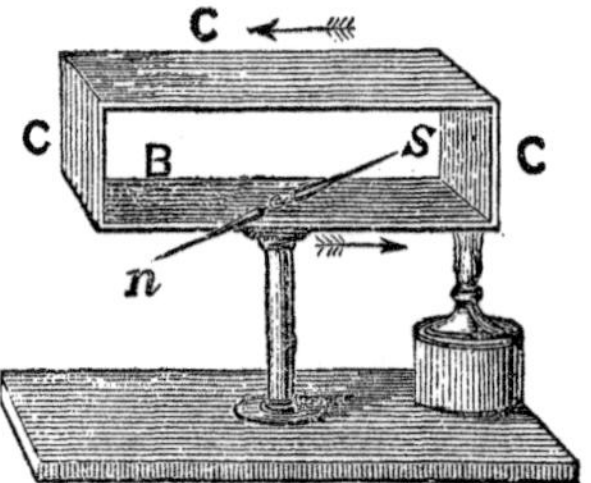

Fig. 12.

Here, then, we have a case in which heat energy goes out of existence, and is converted into that of an electric current, and we may even arrange matters so as to make, on this principle, an instrument which shall be an extremely delicate test of the existence of heat.

By having a number of junctions of bismuth and

antimony, as in Fig. 13, and heating the upper set, while the lower remain cool, we get a strong current going from the bismuth to the antimony across the heated junctions, and we may pass the current so produced round the wire of a galvanometer, and thus, by increasing the number of our junctions, and also by using a very delicate galvanometer, we may get a very perceptible effect for the smallest heating of the upper junctions. This arrangement is called the *thermopile*, and, in conjunction with the reflecting galvanometer, it affords the most delicate means known for detecting small quantities of heat.

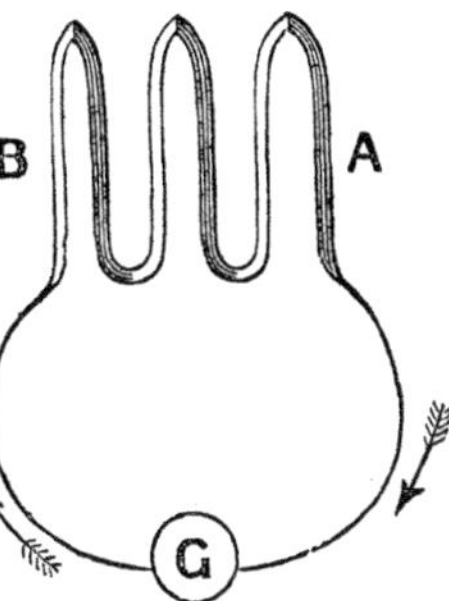

Fig. 13.

162. The last transmutation on our list with respect to absorbed heat is that in which this species of energy is transformed into *radiant light and heat.* This takes place whenever a hot body cools in an open space—the sun, for instance, parts with a large quantity of his heat in this way; and it is due, in part at least, to this process that a hot body cools in air, and wholly to it that such a body cools in vacuo. It is, moreover, due to the penetration of our eye by radiant energy that we are able to see hot bodies, and thus the very fact that we see them implies that they are parting with their heat.

Radiant energy moves through space with the enormous velocity of 188,000 miles in one second. It takes about eight minutes to come from the sun to our earth, so that if our luminary were to be suddenly extinguished, we should have eight minutes, respite before the catastrophe overtook us. Besides the rays that affect the eye, there are others which we cannot see, and which may therefore be termed dark rays. A body, for instance, may not be hot enough to be self-luminous, and yet it may be rapidly cooling and changing its heat into radiant energy, which is given off by the body, even although neither the eye nor the touch may be competent to detect it. It may nevertheless be detected by the thermopile, which was described in Art. 161. We thus see how strong is the likeness between a heated body and a sounding one. For just as a sounding body gives out part of its sound energy to the atmosphere around it, so does a heated body give out part of its heat energy to the ethereal medium around it. When, however, we consider the rates of motion of these energies through their respective media, there is a mighty difference between the two, sound travelling through the air with the velocity of 1100 feet a second, while radiant energy moves over no less a space than 188,000 miles in the same portion of time.

Chemical Separation.

163. We now come to the energy denoted by chemical separation, such as we possess when we have coal or carbon in one place, and oxygen in another. Very evi-

dently this form of energy of position is transmuted into *heat* when we burn the coal, or cause it to combine with the oxygen of the air; and generally, whenever chemical combination takes place, we have the production of heat, even although other circumstances may interfere to prevent its recognition.

Now, in accordance with the principle of conservation, it may be expected that, if a definite quantity of carbon or of hydrogen be burned under given circumstances, there will be a definite production of heat; that is to say, a ton of coals or of coke, when burned, will give us so many heat units, and neither more or less. We may, no doubt, burn our ton in such a way as to economize more or less of the heat produced; but, as far as the mere production of heat is concerned, if the quantity and quality of the material burned and the circumstances of combustion be the same, we expect the same amount of heat.

164. The following table, derived from the researches of Andrews, and those of Favre and Silbermann, shows us how many units of heat we may get by burning a kilogramme of various substances.

UNITS *of* HEAT *developed by* COMBUSTION *in* OXYGEN.

Substance Burned.	Kilogrammes of Water raised 1° C. by the combustion of one kilogramme of each substance.
Hydrogen	34,135
Carbon	7,990
Sulphur	2,263

Substance Burned.	Kilogrammes of Water raised 1° C. by the combustion of one kilogramme of each substance.
Phosphorus	5,747
Zinc	1,301
Iron	1,576
Tin	1,233
Olefiant Gas	11,900
Alcohol	7,016

165. There are other methods, besides combustion, by which chemical combination takes place.

When, for instance, we plunge a piece of metallic iron into a solution of copper, we find that when we take it out, its surface is covered with copper. Part of the iron has been dissolved, taking the place of the copper, which has therefore been thrown, in its metallic state, upon the surface of the iron. Now, in this operation heat is given out—we have in fact burned, or oxidized, the iron, and we are thus furnished with a means of arranging the metals, beginning with that which gives out most heat, when used to displace the metal at the other extremity of the series.

166. The following list has been formed, on this principle, by Dr. Andrews :—

1. Zinc
2. Iron
3. Lead
4. Copper
5. Mercury
6. Silver
7. Platinum

—that is to say, the metal platinum can be displaced by any other metal of the series, but we shall get most heat if we use zinc to displace it.

We may therefore assume that if we displace a definite quantity of platinum by a definite quantity of zinc, we shall get a definite amount of heat. Suppose, however, that instead of performing the operation in one step, we make two of it. Let us, for instance, first of all displace copper by means of zinc, and then platinum by means of copper. Is it not possible that the one of these processes may be more fruitful in heat giving than the other? Now, Andrews has shown us that we cannot gain an advantage over Nature in this way, and that if we use our zinc first of all to displace iron, or copper, or lead, and then use this metal to displace platinum, we shall obtain just the very same amount of heat as if we had used the zinc to displace the platinum at once.

167. It ought here to be mentioned that, very generally, chemical action is accompanied with a change of molecular condition.

A solid, for instance, may be changed into a liquid, or a gas into a liquid. Sometimes the one change counteracts the other as far as apparent heat is concerned; but sometimes, too, they co-operate together to increase the result. Thus, when a gas is absorbed by water, much heat is evolved, and we may suppose the result to be due in part to chemical combination, and in part to the condensation of the gas into a liquid, by which means its latent heat is rendered sensible. On the other hand, when a liquid unites with a solid, or when two solids unite with one another, and the product is a liquid, we have very often the absorption of heat, the heat rendered latent by the dissolution of the solid being more than that generated by combination. Freezing mixtures owe their cooling properties to this cause; thus, if snow and salt be mixed together, they liquefy each other, and the result is brine of a temperature much lower than that of either the ingredients.

168. When heterogeneous metals, such as zinc and copper, are soldered together, we have apparently a conversion of the energy of chemical separation into that of *electrical separation.* This was first suggested by Volta as the origin of the electrical separation which we see in the voltaic current, and recently its existence has been distinctly proved by Sir W. Thomson.

To render manifest this conversion of energy, let us solder a piece of zinc and copper together—if we now test the bar by means of a delicate electrometer we shall find that the zinc is positively, while the copper is negatively, electrified. We have here, therefore, an instance of the transmutation of one form of energy of position into another; so much energy of chemical separation disappearing in order to produce so much electrical separation. This explains the fact recorded in Art. 93, where we saw that if a battery be insulated and its poles kept apart, the one will be charged with positive, and the other with negative, electricity.

169. But further, when such a voltaic battery is in action, we have a transmutation of chemical separation into *electricity in motion.* To see this, let us consider what takes place in such a battery.

Here no doubt the sources of electrical excitement are the points of contact of the zinc and platinum, where, as we see by our last article, we have electrical separation produced. But this of itself would not produce a current, for an electrical current implies very considerable energy, and must be fed by something. Now, in the voltaic battery we have two things which accompany each other, and which are manifestly connected together. In the first place we have the combustion, or at least the oxidation and dissolution, of the zinc; and we have, secondly, the production of a powerful current. Now, evidently, the first of these is that which feeds the second, or, in other words, the energy of chemical separation of the metallic zinc is transmuted into that of an electrical current, the zinc being virtually burned in the process of transmutation.

170. Finally, as far as we are aware, the energy of chemical separation is not directly transmuted into radiant light and heat.

Electrical Separation.

171. In the first place the energy of electrical separation is obviously transmuted into that of *visible motion,* when two oppositely electrified bodies approach each other.

172. Again, it is transmuted into a *current of electricity,* and ultimately into heat, when a spark passes between two oppositely electrified bodies.

It ought, therefore, to be borne in mind that when the flash is seen there is no longer electricity, what we see being merely air, or some other material, intensely heated by the discharge. Thus a man might be rendered insensible by a flash of lightning without his seeing the flash—for the effect of the discharge upon the man, and its effect in heating the air, might be phenomena so nearly simultaneous that the man might become insensible before he could perceive the flash.

Electricity in Motion.

173. This energy is transmuted into that of *visible motion* when two wires conveying electrical currents in the same direction attract each other. When, for instance, two circular currents float on water, both going in the direction of the hands of a watch, we have seen from Art. 100 that they will move towards each other. Now, here there is, in truth, a lessening of the intensity of each current when the motion is taking place, for

we know (Art. 104) that when a circuit is moved into the presence of another circuit conveying a current, there is produced by induction a current in the opposite direction; and hence we perceive that, when two similar currents approach each other, each is diminished by means of this inductive influence—in fact, a certain amount of current energy disappears from existence in order that an equivalent amount of the energy of visible motion may be produced.

174. Electricity in motion is transmuted into *heat* during the passage of a current along a thin wire, or any badly conducting substance—the wire is heated in consequence, and may even become white hot. Most frequently the energy of an electric current is spent in heating the wires and other materials that form the circuit. Now, the energy of such a current is fed by the burning or oxidation of the metal (generally zinc) which is used in the circuit, so that the ultimate effect of this combustion is the heating of the various wires and other materials through which the current passes.

175. We may, in truth, burn or oxidize zinc in two ways—we may oxidize it, as we have just seen, in the voltaic battery, and we shall find that by the combustion of a kilogramme of zinc a definite amount of heat is produced. Or we may oxidize our zinc by dissolving it in acid in a single vessel, when, without going through the intermediate process of a current, we shall get just as much heat out of a kilogramme of zinc as we did in the

former case. In fact, whether we oxidize our zinc by the battery, or in the ordinary way, the quantity of heat produced will always bear the same relation to the quantity of zinc consumed; the only difference being that, in the ordinary way of oxidizing zinc, the heat is generated in the vessel containing the zinc and acid, while in the battery it may make its appearance a thousand miles away, if we have a sufficiently long wire to convey our current.

176. This is, perhaps, the right place for alluding to a discovery of Peltier, that a current of positive electricity passing across a junction of bismuth and antimony in the direction from the bismuth to the antimony appears to produce cold.

To understand the significance of this fact we must consider it in connection with the thermo-electric current, which we have seen, from Art. 161, is established in a circuit of bismuth and antimony, of which one junction is hotter than the other. Suppose we have a circuit of this kind with both its junctions at the temperature of 100° C. to begin with. Suppose, next, that while we protect one junction, we expose the other to the open air—it will, of course, lose heat, so that the protected junction will now be hotter than the other. The consequence will be (Art. 161) that a current of positive electricity will pass along the protected junction from the bismuth to the antimony.

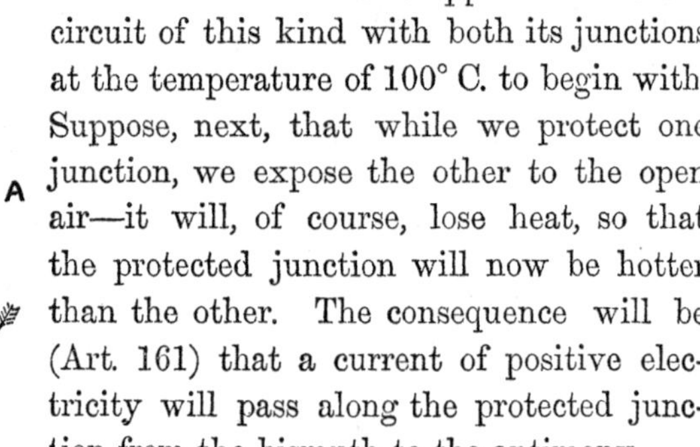

Fig. 14.

Now, here we have an apparent anomaly, for the circuit is cooling—that is to say, it is losing energy—but at the very same time it is manifesting energy in another shape, namely, in that of an electric current, which is circulating round it. Clearly, then, some of the heat of this circuit must be spent in generating this current; in fact, we should expect the circuit to act as a heat engine, only producing current energy instead of mechanical energy, and hence (Art. 152) we should expect to see a conveyance of heat from the hotter to the colder parts of the circuit. Now, this is precisely what the current does, for, passing along the hotter junction, in the direction of the arrow-head, it cools that junction, and heats the colder one at C,—in other words, it carries heat from the hotter to the colder parts of the circuit. We should have been very much surprised had such a current cooled C and heated H, for then we should have had a manifestation of current energy, accompanied with the conveyance of heat from a colder to a hotter substance, which is against the principle of Art. 152.

177. Finally, the energy of electricity in motion is converted into that of *chemical separation*, when a current of electricity is made to decompose a body. Part of the energy of the current is spent in this process, and we shall get so much less heat from it in consequence. Suppose, for instance, that by oxidizing so much zinc in the battery we get, under ordinary circum-

stances, 100 units of heat. Let us, however, set the battery to decompose water, and we shall probably find that by oxidizing the same amount of zinc we get now only 80 units of heat. Clearly, then, the deficiency or 20 units have gone to decompose the water. Now, if we explode the mixed gases which are the result of the decomposition, we shall get back these 20 units of heat precisely, and neither more nor less; and thus we see that amid all such changes the quantity of energy remains the same.

Radiant Energy.

178. This form of energy is converted into *absorbed heat* whenever it falls upon an opaque substance—some of it, however, is generally conveyed away by reflexion, but the remainder is absorbed by the body, and consequently heats it.

It is a curious question to ask what becomes of the radiant light from the sun that is not absorbed either by the planets of our system, or by any of the stars. We can only reply to such a question, that *as far as we can judge from our present knowledge*, the radiant energy that is not absorbed must be conceived to be traversing space at the rate of 188,000 miles a second.

179. There is only one more transmutation of radiant energy that we know of, and that is when it promotes *chemical separation*. Thus, certain rays of the sun are known to have the power of decomposing chloride of

silver, and other chemical compounds. Now, in all such cases there is a transmutation of radiant energy into that of chemical separation. The sun's rays, too, decompose carbonic acid in the leaves of plants, the carbon going to form the woody fibre of the plant, while the oxygen is set free into the air; and of course a certain proportion of the energy of the solar rays is consumed in promoting this change, and we have so much less heating effect in consequence.

But all the solar rays have not this power—for the property of promoting chemical change is confined to the blue and violet rays, and some others which are not visible to the eye. Now, these rays are entirely absent from the radiation of bodies at a comparatively low temperature, such as an ordinary red heat, so that a photographer would find it impossible to obtain the picture of a red-hot body, whose only light was in itself.

180. The actinic, or chemically active, rays of the sun decompose carbonic acid in the leaves of plants, and they disappear in consequence, or are absorbed; this may, therefore, be the reason why very few such rays are either reflected or transmitted from a sun-lit leaf, in consequence of which the photographer finds it difficult to obtain an image of such a leaf; in other words, the rays which would have produced a chemical change on his photographic plate have all been used up by the leaf for peculiar purposes of its own.

181. And here it is important to bear in mind that

while animals in the act of breathing consume the oxygen of the air, turning it into carbonic acid, plants, on the other hand, restore the oxygen to the air; thus the two kingdoms, the animal and the vegetable, work into each other's hands, and the purity of the atmosphere is kept up.

CHAPTER V.

HISTORICAL SKETCH: THE DISSIPATION OF ENERGY.

182. In the last chapter we have endeavoured to exhibit the various transmutations of energy, and, while doing so, to bring forward evidence in favour of the theory of conservation, showing that it enables us to couple together known laws, and also to discover new ones—showing, in fine, that it bears about with it all the marks of a true hypothesis.

It may now, perhaps, be instructive to look back and endeavour to trace the progress of this great conception, from its first beginning among the ancients, up to its triumphant establishment by the labours of Joule and his fellow-workers.

183. Mathematicians inform us that if matter consists of atoms or small parts, which are actuated by forces depending only upon the distances between these parts, and not upon the velocity, then it may be demonstrated that the law of conservation of energy will hold good. Thus we see that conceptions regarding atoms and their

forces are allied to conceptions regarding energy. A medium of some sort pervading space seems also necessary to our theory. In fine, a universe composed of atoms, with some sort of medium between them, is to be regarded as the machine, and the laws of energy as the laws of working of this machine. It may be that a theory of atoms of this sort, with a medium between them, is not after all the simplest, but we are probably not yet prepared for any more general hypothesis. Now, we have only to look to our own solar system, in order to see on a large scale an illustration of this conception, for there we have the various heavenly bodies attracting one another, with forces depending only on the distances between them, and independent of the velocities; and we have likewise a medium of some sort, in virtue of which radiant energy is conveyed from the sun to the earth. Perhaps we shall not greatly err if we regard a molecule as representing on a small scale something analogous to the solar system, while the various atoms which constitute the molecule may be likened to the various bodies of the solar system. The short historical sketch which we are about to give will embrace, therefore, along with energy, the progress of thought and speculation with respect to atoms and also with respect to a medium, inasmuch as these subjects are intimately connected with the doctrines of energy.

Heraclitus on Energy.

184. Heraclitus, who flourished at Ephesus, B.C. 500, declared that fire was the great cause, and that all things were in a perpetual flux. Such an expression will no doubt be regarded as very vague in these days of precise physical statements; and yet it seems clear that Heraclitus must have had a vivid conception of the innate restlessness and energy of the universe, a conception allied in character to, and only less precise than that of modern philosophers, who regard matter as essentially dynamical.

Democritus on Atoms.

185. Democritus, who was born 470 B.C., was the originator of the doctrine of atoms, a doctrine which in the hands of John Dalton has enabled the human mind to lay hold of the laws which regulate chemical changes, as well as to picture to itself what is there taking place. Perhaps there is no doctrine that has nowadays a more intimate connection with the industries of life than this of atoms, and it is probable that no intelligent director of chemical industry among civilized nations fails to picture to his own mind, by means of this doctrine, the inner nature of the changes which he sees with his eyes. Now, it is a curious circumstance that Bacon should have lighted upon this very doctrine of atoms, in order to point one of his philosophical morals.

"Nor is it less an evil" (says he), "that in their philosophies and contemplations men spend their labour in investigating and treating of the first principles of things, and the extreme limits of nature, when all that is useful and of avail in operation is to be found in what is intermediate. Hence it happens that men continue to abstract Nature till they arrive at potential and unformed matter; and again they continue to divide Nature, until they have arrived at the atom; things which, even if true, can be of little use in helping on the fortunes of men."

Surely we ought to learn a lesson from these remarks of the great Father of experimental science, and be very cautious before we dismiss any branch of knowledge or train of thought as essentially unprofitable.

Aristotle on a Medium.

186. As regards the existence of a medium, it is remarked by Whewell that the ancients also caught a glimpse of the idea of a medium, by which the qualities of bodies, as colours and sounds are perceived, and he quotes the following from Aristotle :—

"In a void there could be no difference of up and down; for, as in nothing there are no differences, so there are none in a privation or negation."

Upon this the historian of science remarks, "It is easily seen that such a mode of reasoning elevates the familiar forms of language, and the intellectual connexions of terms, to a supremacy over facts."

Nevertheless, may it not be replied that our conceptions

of matter are deduced from the familiar experience, that certain portions of space affect us in a certain manner; and, consequently, are we not entitled to say there must be something where we experience the difference of up or down? Is there, after all, a very great difference between this argument and that of modern physicists in favour of a plenum, who tell us that matter cannot act where it is not?

Aristotle seems also to have entertained the idea that light is not any body, or the emanation of any body (for that, he says, would be a kind of body), and that therefore light is an energy or act.

The Ideas of the Ancients were not Prolific.

187. These quotations render it evident that the ancients had, in some way, grasped the idea of the essential unrest and energy of things. They had also the idea of small particles or atoms, and, finally, of a medium of some sort. And yet these ideas were not prolific—they gave rise to nothing new.

Now, while the historian of science is unquestionably right in his criticism of the ancients, that their ideas were not distinct and appropriate to the facts, yet we have seen that they were not wholly ignorant of the most profound and deeply-seated principles of the material universe. In the great hymn chanted by Nature, the fundamental notes were early heard, but yet it required long centuries of patient waiting for the practised ear of the skilled musician to appreciate the mighty harmony aright. Or, perhaps, the attempts of the ancients were as the sketches of a child who just contrives to exhibit, in a rude way, the leading outlines of a building; while the conceptions of the practised physicist are more allied to those of the architect, or, at least, of one who has realized, to some extent, the architect's views.

188. The ancients possessed great genius and intellectual power, but they were deficient in physical conceptions, and, in consequence, their ideas were not prolific. It cannot indeed be said that we of the present age are deficient in such conceptions; nevertheless, it may be questioned whether there is not a tendency to rush into the opposite extreme, and to work physical conceptions to an excess. Let us be cautious that in avoiding Scylla, we do not rush into Charybdis. For the universe has more than one point of view, and there are possibly regions which will not yield their treasures to the most determined physicists, armed only with kilogrammes and metres and standard clocks.

Descartes, Newton, and Huyghens on a Medium.

189. In modern times Descartes, author of the vertical hypothesis, necessarily presupposed the existence of a medium in inter-planetary spaces, but on the other hand he was one of the originators of that idea which regards light as a series of particles shot out from a luminous body. Newton likewise conceived the existence of a

medium, although he became an advocate of the theory of emission. It is to Huyghens that the credit belongs of having first conceived the undulatory theory of light with sufficient distinctness to account for double refraction. After him, Young, Fresnel, and their followers, have greatly developed the theory, enabling it to account for the most complicated and wonderful phenomena.

Bacon on Heat.

190. With regard to the nature of heat, Bacon, whatever may be thought of his arguments, seems clearly to have recognized it as a species of motion. He says, "From these instances, viewed together and individually, the nature of which heat is the limitation seems to be motion;" and again he says, "But when we say of motion that it stands in the place of a genus to heat, we mean to convey, not that *heat* generates *motion* or *motion heat* (although even both may be true in some cases), but that essential heat is motion and nothing else."

Nevertheless it required nearly three centuries before the true theory of heat was sufficiently rooted to develop into a productive hypothesis.

Principle of Virtual Velocities.

191. In a previous chapter we have already detailed the labours in respect of heat of Davy, Rumford, and Joule. Galileo and Newton, if they did not grasp the dynamical nature of heat, had yet a clear conception of

the functions of a machine. The former saw that what we gain in power we lose in space; while the latter went further, and saw that a machine, if left to itself, is strictly limited in the amount of work which it can accomplish, although its energy may vary from that of motion to that of position, and back again, according to the geometric laws of the machine.

Rise of true Conceptions regarding Work.

192. There can, we think, be no question that the great development of industrial operations in the present age has indirectly furthered our conceptions regarding work. Humanity invariably strives to escape as much as possible from hard work. In the days of old those who had the power got slaves to work for them; but even then the master had to give some kind of equivalent for the work done. For at the very lowest a slave is a machine, and must be fed, and is moreover apt to prove a very troublesome machine if not properly dealt with. The great improvements in the steam engine, introduced by Watt, have done as much, perhaps, as the abolition of slavery to benefit the working man. The hard work of the world has been put upon iron shoulders, that do not smart; and, in consequence, we have had an immense extension of industry, and a great amelioration in the position of the lower classes of mankind. But if we have transferred our hard work to machines, it is necessary to know how to question a

machine—how to say to it, At what rate can you labour? how much work can you turn out in a day? It is necessary, in fact, to have the clearest possible idea of what work is.

Our readers will see from all this that men are not likely to err in their method of measuring work. The principles of measurement have been stamped as it were with a brand into the very heart and brain of humanity. To the employer of machinery or of human labour, a false method of measuring work simply means ruin; he is likely, therefore, to take the greatest possible pains to arrive at accuracy in his determination.

Perpetual Motion.

193. Now, amid the crowd of workers smarting from the curse of labour, there rises up every now and then an enthusiast, who seeks to escape by means of an artifice from this insupportable tyranny of work. Why not construct a machine that will go on giving you work without limit without the necessity of being fed in any way. Nature must have some weak point in her armour; there must surely be some way of getting round her; she is only tyrannous on the surface, and in order to stimulate our ingenuity, but will yield with pleasure to the persistence of genius.

Now, what can the man of science say to such an enthusiast? He cannot tell him that he is intimately acquainted with all the forces of Nature, and can prove that perpetual motion is impossible; for, in truth, he knows very little of these forces. But he does think that he has entered into the spirit and design of Nature, and therefore he denies at once the possibility of such a machine. But he denies it intelligently, and works out this denial of his into a theory which enables him to discover numerous and valuable relations between the properties of matter—produces, in fact, the laws of energy and the great principle of conservation.

Theory of Conservation.

194. We have thus endeavoured to give a short sketch of the history of energy, including its allied problems, up to the dawn of the strictly scientific period. We have seen that the unfruitfulness of the earlier views was due to a want of scientific clearness in the conceptions entertained, and we have now to say a few words regarding the theory of conservation.

Here also the way was pointed out by two philosophers, namely, Grove in this country, and Mayer on the continent, who showed certain relations between the various forms of energy; the name of Séguin ought likewise to be mentioned. Nevertheless, to Joule belongs the honour of establishing the theory on an incontrovertible basis: for, indeed, this is pre-eminently a case where speculation has to be tested by unimpeachable experimental evidence. Here the magnitude of the principle is so vast, and its importance is so great, that it requires the strong fire of genius, joined to the patient labours of the scientific experimentalist, to forge the rough ore into a good weapon that will cleave its way through all obstacles into the very citadel of Nature, and into her most secret recesses.

Following closely upon the labours of Joule, we have those of William and James Thomson, Helmholtz, Rankine, Clausius, Tait, Andrews, Maxwell, who, along with many others, have advanced the subject; and while Joule gave his chief attention to the laws which regulate the transmutation of mechanical energy into heat, Thomson, Rankine, and Clausius gave theirs to the converse problem, or that which relates to the transmutation of heat into mechanical energy. Thomson, especially, has pushed forward so resolutely from this point of view that he has succeeded in grasping a principle scarcely inferior in importance to that of the conservation of energy itself, and of this principle it behoves us now to speak.

Dissipation of Energy.

195. Joule, we have said, proved the law according to which work may be changed into heat; and Thomson and others, that according to which heat may be changed into work. Now, it occurred to Thomson that there was a very important and significant difference between these two laws, consisting in the fact that, while you can with the greatest ease transform work into heat, you can by no method in your power transform all the heat back

again into work. In fact, the process is not a reversible one; and the consequence is that the mechanical energy of the universe is becoming every day more and more changed into heat.

It is easily seen that if the process were reversible, one form of a perpetual motion would not be impossible. For, without attempting to create energy by a machine, all that would be needed for a perpetual motion would be the means of utilizing the vast stores of heat that lie in all the substances around us, and converting them into work. The work would no doubt, by means of friction and otherwise, be ultimately reconverted into heat; but if the process be reversible, the heat could again be converted into work, and so on for ever. But the irreversibility of the process puts a stop to all this. In fact, I may convince myself by rubbing a metal button on a piece of wood how easily work can be converted into heat, while the mind completely fails to suggest any method by which this heat can be reconverted into work.

Now, if this process goes on, and always in one direction, there can be no doubt about the issue. The mechanical energy of the universe will be more and more transformed into universally diffused heat, until the universe will no longer be a fit abode for living beings.

The conclusion is a startling one, and, in order to bring it more vividly before our readers, let us now proceed to acquaint ourselves with the various forms of useful energy that are at present at our disposal, and at the same time endeavour to trace the ultimate sources of these supplies.

Natural Energies and their Sources.

196. Of energy in repose we have the following varieties :—(1.) The energy of fuel. (2.) That of food. (3.) That of a head of water. (4.) That which may be derived from the tides. (5.) The energy of chemical separation implied in native sulphur, native iron, &c.

Then, with regard to energy in action, we have mainly the following varieties :—

(1.) The energy of air in motion. (2.) That of water in motion.

Fuel.

197. Let us begin first with the energy implied in fuel. We can, of course, burn fuel, or cause it to combine with the oxygen of the air ; and we are thereby provided with large quantities of heat of high temperature, by means of which we may not only warm ourselves and cook our food, but also drive our heat-engines, using it, in fact, as a source of mechanical power.

Fuel is of two varieties—wood and coal. Now, if we consider the origin of these we shall see that they are produced by the sun's rays. Certain of these rays, as we have already remarked (Art. 180), decompose carbonic acid in the leaves of plants, setting free the oxygen, while the carbon is used for the structure or wood of the plant. Now, the energy of these rays is spent in this process, and, indeed, there is not enough of such energy left to produce a good photographic impression of the leaf of a plant, because it is all spent in making wood.

We thus see that the energy implied in wood is derived from the sun's rays, and the same remark applies to coal. Indeed, the only difference between wood and coal is one of age: wood being recently turned out from Nature's laboratory, while thousands of years have elapsed since coal formed the leaves of living plants.

198. We are, therefore, perfectly justified in saying that the energy of fuel is derived from the sun's rays; * coal being the store which Nature has laid up as a species of capital for us, while wood is our precarious yearly income.

We are thus at present very much in the position of a young heir, who has only recently come into his estate, and who, not content with the income, is rapidly squandering his realized property. This subject has been forcibly brought before us by Professor Jevons, who has remarked that not only are we spending our capital, but we are spending the most available and valuable part of it. For we are now using the surface coal; but a time will come when this will be exhausted, and we shall be compelled to go deep down for our

* This fact seems to have been known at a comparatively early period to Herschel and the elder Stephenson.

supplies. Now, regarded as a source of energy, such supplies, if far down, will be less effective, for we have to deduct the amount of energy requisite in order to bring them to the surface. The result is that we must contemplate a time, however far distant, when our supplies of coal will be exhausted, and we shall be compelled to resort to other sources of energy.

Food.

199. The energy of food is analogous to that of fuel, and serves similar purposes. For just as fuel may be used either for producing heat or for doing work, so food has a twofold office to perform. In the first place, by its gradual oxidation, it keeps up the temperature of the body; and in the next place it is used as a source of energy, on which to draw for the performance of work. Thus a man or a horse that works a great deal requires to eat more food than if he does not work at all. Thus, also, a prisoner condemned to hard labour requires a better diet than one who does not work, and a soldier during the fatigues of war finds it necessary to eat more than during a time of peace.

Our food may be either of animal or vegetable origin—if it be the latter, it is immediately derived, like fuel, from the energy of the sun's rays; but if it be the former, the only difference is that it has passed through the body of an animal before coming to us: the animal has eaten grass, and we have eaten the animal.

In fact, we make use of the animal not only as a variety of nutritious food, but also to enable us indirectly to utilize those vegetable products, such as grasses, which we could not make use of directly with our present digestive organs.

Head of Water.

200. The energy of a head of water, like that of fuel and food, is brought about by the sun's rays. For the sun vaporizes the water, which, condensed again in upland districts, becomes available as a head of water.

There is, however, the difference that fuel and food are due to the actinic power of the sun's rays, while the evaporation and condensation of water are caused rather by their heating effect.

Tidal Energy.

201. The energy derived from the tides has, however, a different origin. In Art. 133 we have endeavoured to show how the moon acts upon the fluid portions of our globe, the result of this action being a very gradual stoppage of the energy of rotation of the earth.

It is, therefore, to this motion of rotation that we must look as the origin of any available energy derived from tidal mills.

Native Sulphur, &c.

202. The last variety of available energy of position in our list is that implied in native sulphur, native iron, &c. It has been remarked by Professor Tait, to whom this method of reviewing our forces is due, that this may be the primeval form of energy, and that the interior of the earth may, as far as we know, be wholly composed of matter in its uncombined form. As a source of available energy it is, however, of no practical importance.

Air and Water in Motion.

203. We proceed next to those varieties of available energy which represent motion, the chief of which are air in motion and water in motion. It is owing to the former that the mariner spreads his sail, and carries his vessel from one part of the earth's surface to another, and it is likewise owing to the same influence that the windmill grinds our corn. Again, water in motion is used perhaps even more frequently than air in motion as a source of motive power.

Both these varieties of energy are due without doubt to the heating effect of the sun's rays. We may, therefore, affirm that with the exception of the totally insignificant supply of native sulphur, &c., and the small number of tidal mills which may be in operation, all our available energy is due to the sun.

The Sun—a Source of High Temperature Heat.

204. Let us, therefore, now for a moment direct our attention to that most wonderful source of energy, the Sun.

We have here a vast reservoir of high temperature heat; now, this is a kind of superior energy which has always been in much request. Numberless attempts have been made to construct a perpetual light, just as similar attempts have been made to construct a perpetual motion, with this difference, that a perpetual light was supposed to result from magical powers, while a perpetual motion was attributed to mechanical skill.

Sir Walter Scott alludes to this belief in his description of the grave of Michael Scott, which is made to contain a perpetual light. Thus the Monk who buried the wizard tells William of Deloraine—

> "Lo, Warrior! now the Cross of Red
> Points to the Grave of the mighty dead;
> Within it burns a wondrous light,
> To chase the spirits that love the night.
> That lamp shall burn unquenchably
> Until the eternal doom shall be."

And again, when the tomb was opened, we read—

> "I would you had been there to see
> How the light broke forth so gloriously,
> Stream'd upward to the chancel roof,
> And through the galleries far aloof!
> No earthly flame blazed e'er so bright."

No earthly flame—there the poet was right—certainly not of this earth, where light and all other forms of superior energy are essentially evanescent.

A Perpetual Light Impossible.

205. In truth, our readers will at once perceive that a perpetual light is only another name for a perpetual motion, because we can always derive visible energy out of high temperature heat—indeed, we do so every day in our steam engines.

When, therefore, we burn coal, and cause it to combine with the oxygen of the air, we derive from the process a large amount of high temperature heat. But is it not possible, our readers may ask, to take the carbonic acid which results from the combustion, and by means of low temperature heat, of which we have always abundance at our disposal, change it back again into carbon and oxygen? All this would be possible if what may be termed the temperature of disassociation—that is to say, the temperature at which carbonic acid separates into its constituents—were a low temperature, and it would also be possible if rays from a source of low temperature possessed sufficient actinic power to decompose carbonic acid.

But neither of these is the case. Nature will not be caught in a trap of this kind. As if for the very purpose of stopping all such speculations, the temperatures of disassociation for such substances as carbonic acid are very high, and the actinic rays capable of causing their decomposition belong only to sources of exceedingly high temperature, such as the sun.*

Is the Sun an Exception?

206. We may, therefore, take it for granted that a perpetual light, like a perpetual motion, is an impossibility; and we have then to inquire if the same argument applies to our sun, or if an exception is to be made in his favour. Does the sun stand upon a footing of his own, or is it merely a question of time with him, as with all other instances of high temperature heat? Before attempting to answer this question let us inquire into the probable origin of the sun's heat.

Origin of the Sun's Heat.

207. Now, some might be disposed to cut the Gordian knot of such an inquiry by asserting that our luminary was at first created hot; yet the scientific mind finds itself disinclined to repose upon such an assertion. We pick up a round pebble from the beach, and at once acknowledge there has been some physical cause for the shape into which it has been worn. And so with regard to the heat of the sun, we must ask ourselves if there be not some cause not wholly imaginary, but one which we know, or at least suspect, to be perhaps still in operation, which can account for the heat of the sun.

Now, here it is more easy to show what cannot

* This remark is due to Sir William Thomson.

account for the sun's heat than what can do so. We may, for instance, be perfectly certain that it cannot have been caused by chemical action. The most probable theory is that which was first worked out by Helmholtz and Thomson;* and which attributes the heat of the sun to the primeval energy of position possessed by its particles. In other words, it is supposed that these particles originally existed at a great distance from each other, and that, being endowed with the force of gravitation, they have since gradually come together, while in this process heat has been generated just as it would be if a stone were dropped from the top of a cliff towards the earth.

208. Nor is this case wholly imaginary, but we have some reason for thinking that it may still be in operation in the case of certain nebulæ which, both in their constitution as revealed by the spectroscope, and in their general appearance, impress the beholder with the idea that they are not yet fully condensed into their ultimate shape and size.

If we allow that by this means our luminary has obtained his wonderful store of high-class energy, we have yet to inquire to what extent this operation is going on at the present moment. Is it only a thing of the past, or is it a thing also of the present? I think we may reply that the sun cannot be condensing very fast, at least, within historical times. For if the

* Mayer and Waterston seem first to have caught the rudiments of this idea.

sun were sensibly larger than at present his total eclipse by the moon would be impossible. Now, such eclipses have taken place, at any rate, for several thousands of years. Doubtless a small army of meteors may be falling into our luminary, which would by this fall tend to augment his heat; yet the supply derived from this source must surely be insignificant. But if the sun be not at present condensing so fast as to derive any sufficient heat from this process, and if his energy be very sparingly recruited from without, it necessarily follows that he is in the position of a man whose expenditure exceeds his income. He is living upon his capital, and is destined to share the fate of all who act in a similar manner. We must, therefore, contemplate a future period when he will be poorer in energy than he is at present, and a period still further in the future when he will altogether cease to shine.

Probable Fate of the Universe.

209. If this be the fate of the high temperature energy of the universe, let us think for a moment what will happen to its visible energy. We have spoken already about a medium pervading space, the office of which appears to be to degrade and ultimately extinguish all differential motion, just as it tends to reduce and ultimately equalize all difference of temperature. Thus the universe would ultimately become an equally heated mass, utterly worthless as far as the production of work is concerned, since such production depends upon difference of temperature.

Although, therefore, in a strictly mechanical sense, there is a conservation of energy, yet, as regards usefulness or fitness for living beings, the energy of the universe is in process of deterioration. Universally diffused heat forms what we may call the great waste-heap of the universe, and this is growing larger year by year. At present it does not sensibly obtrude itself, but who knows that the time may not arrive when we shall be practically conscious of its growing bigness ?

210. It will be seen that in this chapter we have regarded the universe, not as a collection of matter, but rather as an energetic agent—in fact, as a lamp. Now, it has been well pointed out by Thomson, that looked at in this light, the universe is a system that had a beginning and must have an end; for a process of degradation cannot be eternal. If we could view the universe as a candle not lit, then it is perhaps conceivable to regard it as having been always in existence; but if we regard it rather as a candle that has been lit, we become absolutely certain that it cannot have been burning from eternity, and that a time will come when it will cease to burn. We are led to look to a beginning in which the particles of matter were in a diffuse chaotic state, but endowed with the power of gravitation, and we are led to look to an end in which the whole universe will be one equally heated inert mass, and from which everything like life or motion or beauty will have utterly gone away.

CHAPTER VI.

THE POSITION OF LIFE.

211. WE have hitherto confined ourselves almost entirely to a discussion of the laws of energy, as these affect inanimate matter, and have taken little or no account of the position of life. We have been content very much to remain spectators of the contest, apparently forgetful that we are at all concerned in the issue. But the conflict is not one which admits of on-lookers,—it is a universal conflict in which we must all take our share. It may not, therefore, be amiss if we endeavour to ascertain, as well as we can, our true position.

Twofold nature of Equilibrium.

212. One of our earliest mechanical lessons is on the twofold nature of equilibrium. We are told that this may be of two kinds, *stable* and *unstable,* and a very good illustration of these two kinds is furnished by an egg. Let us take a smooth level table, and place an egg upon it; we all know in what manner the egg will lie

on the table. It will remain at rest, that is to say, it will be in equilibrium; and not only so, but it will be in stable equilibrium. To prove this, let us try to displace it with our finger, and we shall find that when we remove the pressure the egg will speedily return to its previous position, and will come to rest after one or two oscillations. Furthermore, it has required a sensible expenditure of energy to displace the egg. All this we express by saying that the egg is in stable equilibrium.

Mechanical Instability.

213. And now let us try to balance the egg upon its longer axis. Probably, a sufficient amount of care will enable us to achieve this also. But the operation is a difficult one, and requires great delicacy of touch, and even after we have succeeded we do not know how long our success may last. The slightest impulse from without, the merest breath of air, may be sufficient to overturn the egg, which is now most evidently in unstable equilibrium. If the egg be thus balanced at the very edge of the table, it is quite probable that in a few minutes it may topple over upon the floor; it is what we may call *an even chance* whether it will do so, or merely fall upon the table. Not that mere chance has anything to do with it, or that its movements are without a cause, but we mean that its movements are decided by some external impulse so exceedingly small as to be utterly beyond our powers of observation. In fact, before making the trial

we have carefully removed everything like a current of air, or want of level, or external impulse of any kind, so that when the egg falls we are completely unable to assign the origin of the impulse that has caused it to do so.

214. Now, if the egg happens to fall over the table upon the floor, there is a somewhat considerable transmutation of energy; for the energy of position of the egg, due to the height which it occupied on the table, has all at once been changed into energy of motion, in the first place, and into heat in the second, when the egg comes into contact with the floor.

If, however, the egg happens to fall upon the table, the transmutation of energy is comparatively small.

It thus appears that it depends upon some external impulse, so infinitesimally small as to elude our observation, whether the egg shall fall upon the floor and give rise to a comparatively large transmutation of energy, or whether it shall fall upon the table and give rise to a transmutation comparatively small.

Chemical Instability.

215. We thus see that a body, or system, in unstable equilibrium may become subject to a very considerable transmutation of energy, arising out of a very small cause, or antecedent. In the case now mentioned, the force is that of gravitation, the arrangement being one of visible mechanical instability. But we may have a sub-

stance, or system, in which the force at work is not gravity, but chemical affinity, and the substance, or system, may, under certain peculiar conditions, become *chemically unstable.*

When a substance is chemically unstable, it means that the slightest impulse of any kind may determine a chemical change, just as in the case of the egg the slightest impulse from without occcasioned a mechanical displacement.

In fine, a substance, or system, chemically unstable bears a relation to chemical affinity somewhat similar to that which a mechanically unstable system bears to gravity. Gunpowder is a familiar instance of a chemically unstable substance. Here the slightest spark may prove the precursor of a sudden chemical change, accompanied by the instantaneous and violent generation of a vast volume of heated gas. The various explosive compounds, such as gun-cotton, nitro-glycerine, the fulminates, and many more, are all instances of structures which are chemically unstable.

Machines are of two kinds.

216. When we speak of a structure, or a machine, or a system, we simply mean a number of individual particles associated together in producing some definite result. Thus, the solar system, a timepiece, a rifle, are examples of inanimate machines; while an animal, a human being, an army, are examples of animated struc-

tures or machines. Now, such machines or structures are of two kinds, which differ from one another not only in the object sought, but also in the means of attaining that object.

217. In the first place, we have structures or machines in which systematic action is the object aimed at, and in which all the arrangements are of a conservative nature, the element of instability being avoided as much as possible. The solar system, a timepiece, a steam-engine at work, are examples of such machines, and the characteristic of all such is their *calculability.* Thus the skilled astronomer can tell, with the utmost precision, in what place the moon or the planet Venus will be found this time next year. Or again, the excellence of a timepiece consists in its various hands pointing accurately in a certain direction after a certain interval of time. In like manner we may safely count upon a steamship making so many knots an hour, at least while the outward conditions remain the same. In all these cases we make our calculations, and we are not deceived—the end sought is regularity of action, and the means employed is a stable arrangement of the forces of nature.

218. Now, the characteristics of the other class of machines are precisely the reverse.

Here the object aimed at is not a regular, but a sudden and violent transmutation of energy, while the means employed are unstable arrangements of natural forces.

A rifle at full cock, with a delicate hair-trigger, is a very good instance of such a machine, where the slightest touch from without may bring about the explosion of the gunpowder, and the propulsion of the ball with a very great velocity. Now, such machines are eminently characterized by their *incalculability.*

219. To make our meaning clear, let us suppose that two sportsmen go out hunting together, each with a good rifle and a good pocket chronometer. After a hard day's work, the one turns to his companion and says:—"It is now six o'clock by my watch; we had better rest ourselves," upon which the other looks at his watch, and he would be very much surprised and exceedingly indignant with the maker, if he did not find it six o'clock also. Their chronometers are evidently in the same state, and have been doing the same thing; but what about their rifles? Given the condition of the one rifle, is it possible by any refinement of calculation to deduce that of the other? We feel at once that the bare supposition is ridiculous.

220. It is thus apparent that, as regards energy, structures are of two kinds. In one of these, the object sought is regularity of action, and the means employed, a stable arrangement of natural forces: while in the other, the end sought is freedom of action, and a sudden transmutation of energy, the means employed being an unstable arrangement of natural forces.

The one set of machines are characterized by their

calculability—the other by their incalculability. The one set, when at work, are not easily put wrong, while the other set are characterized by great delicacy of construction.

An Animal is a delicately-constructed Machine.

221. But perhaps the reader may object to our use of the rifle as an illustration.

For although it is undoubtedly a delicately-constructed machine, yet a rifle does not represent the same surpassing delicacy as that, for instance, which characterizes an egg balanced on its longer axis. Even if at full cock, and with a hair trigger, we may be perfectly certain it will not go off of its own accord. Although its object is to produce a sudden and violent transmutation of energy, yet this requires to be preceded by the application of an amount of energy, however small, to the trigger, and if this be not spent upon the rifle, it will not go off. There is, no doubt, delicacy of construction, but this has not risen to the height of incalculability, and it is only when in the hands of the sportsman that it becomes a machine upon the condition of which we cannot calculate.

Now, in making this remark, we define the position of the sportsman himself in the Universe of Energy.

The rifle is delicately constructed, but not surpassingly so; but sportsman and rifle, together, form a machine of surpassing delicacy, *ergo* the sportsman himself is such a machine. We thus begin to perceive that a human being, or indeed an animal of any kind, is in truth a machine of a delicacy that is practically infinite, the condition or motions of which we are utterly unable to predict.

In truth, is there not a transparent absurdity in the very thought that a man may become able to calculate his own movements, or even those of his fellow?

Life is like the Commander of an Army

222. Let us now introduce another analogy—let us suppose that a war is being carried on by a vast army, at the head of which there is a very great commander. Now, this commander knows too well to expose his person; in truth, he is never seen by any of his subordinates. He remains at work in a well-guarded room, from which telegraphic wires lead to the headquarters of the various divisions. He can thus, by means of these wires, transmit his orders to the generals of these divisions, and by the same means receive back information as to the condition of each.

Thus his headquarters become a centre, into which all information is poured, and out of which all commands are issued.

Now, that mysterious thing called life, about the nature of which we know so little, is probably not unlike such a commander. Life is not a bully, who swaggers out into the open universe, upsetting the laws of energy in all directions, but rather a consummate strategist, who, sitting in his secret chamber, before his wires, directs the movements of a great army.*

223. Let us next suppose that our imaginary army is in rapid march, and let us try to find out the cause of this movement. We find that, in the first place, orders to march have been issued to the troops under them by the commanders of each regiment. In the next place, we learn that staff officers, attached to the generals of the various divisions, have conveyed these orders to the regimental commanders; and, finally, we learn that the order to march has been telegraphed from headquarters to these various generals.

Descending now to ourselves, it is probably somewhere in the mysterious and well-guarded brain-chamber that the delicate directive touch is given which determines our movements. This chamber forms, as it were, the headquarters of the general in command, who is so well withdrawn as to be absolutely invisible to all his subordinates.

224. Joule, Carpenter, and Mayer were at an early period aware of the restrictions under which animals are placed by the laws of energy, and in virtue of which the power of an animal, as far as energy is concerned, is not creative, but only directive. It was seen that, in order

* *See* an article on "The Position of Life," by the author of this work, in conjunction with Mr. J. N. Lockyer, "Macmillan's Magazine," September, 1868; also a lecture on "The Recent Developments of Cosmical Physics," by the author of this work.

to do work, an animal must be fed; and, even at a still earlier period, Count Rumford remarked that a ton of hay will be administered more economically by feeding a horse with it, and then getting work out of the horse, than by burning it as fuel in an engine.

225. In this chapter, the same line of thought has been carried out a little further. We have seen that life is associated with delicately-constructed machines, so that whenever a transmutation of energy is brought about by a living being, could we trace the event back, we should find that the physical antecedent was probably a much less transmutation, while again the antecedent of this would probably be found still less, and so on, as far as we could trace it.

226. But with all this, we do not pretend to have discovered the true nature of life itself, or even the true nature of its relation to the material universe.

What we have ventured is the assertion that, as far as we can judge, life is always associated with machinery of a certain kind, in virtue of which an extremely delicate directive touch is ultimately magnified into a very considerable transmutation of energy. Indeed, we can hardly imagine the freedom of motion implied in life to exist apart from machinery possessed of very great delicacy of construction.

In fine, we have not succeeded in solving the problem as to the true nature of life, but have only driven the difficulty into a borderland of thick darkness, into which the light of knowledge has not yet been able to penetrate.

Organized Tissues are subject to Decay.

227. We have thus learned two things, for, in the first place, we have learned that life is associated with delicacy of construction, and in the next (Art. 220), that delicacy of construction implies an unstable arrangement of natural forces. We have now to remark that the particular force which is thus used by living beings is chemical affinity. Our bodies are, in truth, examples of an unstable arrangement of chemical forces, and the materials which composed them, if not liable to sudden explosion, like fulminating powder, are yet pre-eminently the subjects of decay.

228. Now, this is more than a mere general statement; it is a truth that admits of degrees, and in virtue of which those parts of our bodies which have, during life, the noblest and most delicate office to perform, are the very first to perish when life is extinct.

> "Oh! o'er the eye death most exerts his might,
> And hurls the spirit from her throne of light;
> Sinks those blue orbs in their long last eclipse,
> But spares as yet the charm around the lips."

So speaks the poet, and we have here an aspect of things in which the lament of the poet becomes the true interpretation of nature.

Difference between Animals and Inanimate Machines.

229. We are now able to recognize the difference between the relations to energy of a living being, such as man, and a machine, such as a steam-engine.

There are many points in common between the two. Both require to be fed, and in both there is the transmutation of the energy of chemical separation implied in fuel and food into that of heat and visible motion.

But while the one—the engine—requires for its maintenance only carbon, or some other variety of chemical separation, the other—the living being—demands to be supplied with organized tissue. In fact, that delicacy of construction which is so essential to our well-being, is not something which we can elaborate internally in our own frames—all that we can do is to appropriate and assimilate that which comes to us from without; it is already present in the food which we eat.

Ultimate Dependence of Life upon the Sun.

230. We have already (Art. 203) been led to recognize the sun as the ultimate material source of all the energy which we possess, and we must now regard him as the source likewise of all our delicacy of construction. It requires the energy of his high temperature rays so to wield and manipulate the powerful forces of chemical affinity; so to balance these various forces against each

other, as to produce in the vegetable something which will afford our frames, not only energy, but also delicacy of construction.

Low temperature heat would be utterly unable to accomplish this; it consists of ethereal vibrations which are not sufficiently rapid, and of waves that are not sufficiently short, for the purpose of shaking asunder the constituents of compound molecules.

231. It thus appears that animals are, in more ways than one, pensioners upon the sun's bounty; and those instances, which at first sight appear to be exceptions, will, if studied sufficiently, only serve to confirm the rule.

Thus the recent researches of Dr. Carpenter and Professor Wyville Thomson have disclosed to us the existence of minute living beings in the deepest parts of the ocean, into which we may be almost sure no solar ray can penetrate. How, then, do these minute creatures obtain that energy and delicacy of construction without which they cannot live? in other words, how are they fed?

Now, the same naturalists who discovered the existence of these creatures, have recently furnished us with a very probable explanation of the mystery. They think it highly probable that the whole ocean contains in it organic matter to a very small but yet perceptible extent, forming, as they express it, a soft of diluted soup, which thus becomes the food of these minute creatures.

232. In conclusion, we are dependent upon the sun and centre of our system, not only for the mere energy of our

frames, but also for our delicacy of construction—the future of our race depends upon the sun's future. But we have seen that the sun must have had a beginning, and that he will have an end.

We are thus induced to generalize still further, and regard, not only our own system, but the whole material universe when viewed with respect to serviceable energy, as essentially evanescent, and as embracing a succession of physical events which cannot go on for ever as they are.

But here at length we come to matters beyond our grasp; for physical science cannot inform us what must have been before the beginning, nor yet can it tell us what will take place after the end.

APPENDIX.

CORRELATION OF VITAL WITH CHEMICAL AND PHYSICAL FORCES.

BY JOSEPH LE CONTE,

PROFESSOR OF GEOLOGY AND NATURAL HISTORY IN THE UNIVERSITY OF CALIFORNIA.

CORRELATION OF VITAL WITH CHEMICAL AND PHYSICAL FORCES.

VITAL force; whence is it derived? What is its relation to the other forces of Nature? The answer of modern science to these questions is: It is derived from the lower forces of Nature; it is related to other forces much as these are related to each other—it is correlated with chemical and physical forces.

At one time matter was supposed to be destructible. By combustion or by evaporation matter seemed to be consumed—to pass out of existence; but now we know it only changes its form from the solid or liquid to the gaseous condition—from the visible to the invisible—and that, amid all these changes, the same quantity of matter remains. Creation or destruction of matter, increase or diminution of matter, lies beyond the domain of Science; her domain is confined entirely to the changes of matter. Now, it is the doctrine of modern science that the same is true of force. Force seems often to be annihilated. Two cannon-balls of equal size and velocity meet each other and fall motionless. The immense energy of these moving bodies seems to pass out of existence. But not so; it is changed into heat, and the exact amount of heat may be calculated; moreover, an equal amount of heat may be changed back again into an equal amount of momentum. Here, therefore, force is not lost, but is changed from a visible to an invisible form. Motion is changed from bodily motion into molecular motion. Thus heat, light, electricity, magnetism, chemical affinity, and mechanical force, are transmutable into each other, back and forth; but, amid all these changes, the amount of force remains unchanged. Force is incapàble of destruction, except by the same power which created it. The domain of Science lies within the limits of these changes—creation and annihilation lie outside of her domain.

The mutual convertibility of forces into each other is called *correlation of forces;* the persistence of the same amount, amid all these protean forms, is called *conservation of force.**

* In recent works the word *energy* is used to designate active or working force as distinguished from passive or non-working force. It is in this working condition only that force is conserved, and therefore *conservation of energy* is the proper expression. Nevertheless, since the distinction between force and energy is imperfectly or not at all defined in the higher forms of force, and especially in the domain of life, I have preferred in this article to use the word *force* in the general sense usual until recently.

The correlation of physical forces with each other and with chemical force is now universally acknowledged and somewhat clearly conceived. The correlation of vital force with these is not universally acknowledged, and, where acknowledged, is only imperfectly conceived. In 1859 I published a paper * in which I attempted to put the idea of correlation of vital force with chemical and physical forces in a more definite and scientific form. The views expressed in that paper have been generally adopted by physiologists. Since the publication of the paper referred to, the subject has lain in my mind, and grown at least somewhat. I propose, therefore, now to reëmbody my views in a more popular form, with such additions as have occurred to me since.

There are four planes of material existence, which may be represented as raised one above another. These are: 1. The plane of elementary existence; 2. The plane of chemical compounds, or mineral kingdom; 3. The plane of vegetable existence; and, 4. The plane of animal existence. Their relations to each other are truly expressed by writing them one above the other, thus:

I may sometimes use the word energy instead. If any one should charge me with want of precision in language, my answer is: Our language cannot be more precise until our ideas in this department are far clearer than now.

* *American Journal of Science*, November, 1859. *Philadelphia Magazine*, vol. xix., p. 133.

4. *Animal Kingdom.*
3. *Vegetable Kingdom.*
2. *Mineral Kingdom.*
1. *Elements.*

Now, it is a remarkable fact that there is a special force, whose function it is to raise matter from each plane to the plane above, and to execute movements on the latter. Thus, it is the function of chemical affinity alone to raise matter from No. 1 to No. 2, as well as to execute all the movements, back and forth, by action and reaction; in a word, to produce all the phenomena on No. 2 which together constitute the science of chemistry. It is the prerogative of vegetable life-force alone to lift matter from No. 2 to No. 3, as well as to execute all the movements on that plane, which together constitute the science of vegetable physiology. It is the prerogative of animal life-force alone to lift matter from No. 3 to No. 4, and to preside over the movements on this plane, which together constitute the science of animal physiology. But there is no force in Nature capable of raising matter at once from No. 1 to No. 3, or from No. 2 to No. 4, without stopping and receiving an accession of force, of a different kind, on the intermediate plane. Plants cannot feed upon elements, but only on chemical compounds; animals cannot feed on minerals, but only on vegetables. We shall see in the sequel that this is the necessary result of the principle of conservation of force in vital phenomena.

It is well known that atoms, in a nascent state—i. e., at the moment of their separation from previous combination—are endowed with peculiar and powerful affinity. Oxygen and nitrogen, nitrogen and hydrogen, hydrogen and carbon, which show no affinity for each other under ordinary circumstances, readily unite when one or both are in a nascent condition. The reason seems to be that, when the elements of a compound are torn asunder, the chemical affinity which previously bound them together is set free, ready and eager to unite the nascent elements with whatever they come in contact with. This state of exalted chemical energy is retained but a little while, because it is liable to be changed into some other form of force, probably heat, and is therefore no longer chemical energy. To illustrate by the planes: matter falling down from No. 2 to No. 1 generates force by which matter is lifted from No. 1 to No. 2. Decomposition generates the force by which combination is effected. This principle underlies every thing I shall further say.

There are, therefore, two ideas or principles underlying this paper: 1. The correlation of vital with physical and chemical forces; 2. That in all cases *vital force is produced by decomposition*—is transformed nascent affinity. Neither of these is new. Grove, many years ago, brought out, in a vague manner, the idea that vital force was correlated with chemical and physical forces.* In 1848 Dr. Freke, M. R. I. A., of Dublin, first advanced the idea that vital force of animal life was generated by decomposition. In 1851 the same idea was brought out again by Dr. Watters, of St. Louis. These papers were unknown to me when I wrote my article. They have been sent to me in the last few years by their respective authors. Neither of these authors, however, extends this principle to vegetation, the most fundamental and most important phenomenon of life. In 1857 the same idea was again brought out by Prof. Henry, of the Smithsonian Institution, and by him extended to vegetation. I do not, therefore, now claim to have first advanced this idea, but I do claim to have in some measure rescued it from vagueness, and given it a clearer and more scientific form.

I wish now to apply these principles in the explanation of the most important phenomena of vegetable and animal life:

1. Vegetation.—The most important phenomenon in the life-history of a plant—in fact, the starting-point of all life, both vegetable and animal—is the formation of organic matter in the leaves. The necessary conditions for this wonderful change of mineral into organic mat-

* In 1845 Dr. J. R. Mayer published a paper on "Organic Motion and Nutrition." I have not seen it.

ter seem to be, sunlight, chlorophyl, and living protoplasm, or bioplasm. This is the phenomenon I wish now to discuss.

The plastic matters of which vegetable structure is built are of two kinds—amyloids and albuminoids. The amyloids, or starch and sugar groups, consist of C, H, and O; the albuminoids of C, H, O, N, and a little S and P. The quantity of sulphur and phosphorus is very small, and we will neglect them in this discussion. The food out of which these substances are elaborated are, CO_2, H_2O, and H_3N—carbonic acid, water, and ammonia. Now, by the agency of sunlight in the presence of chlorophyl and bioplasm, these chemical compounds (CO_2, H_2O, H_3N) are torn asunder, or shaken asunder, or decomposed; the excess of O, or of O and H, is rejected, and the remaining elements in a nascent condition combine to form organic matter. To form the amyloids—starch, dextrine, sugar, cellulose—only CO_2 and H_2O are decomposed, and excess of O rejected. To form albuminoids, or protoplasm, CO_2, H_2O, and H_3N, are decomposed, and excess of O and H rejected.

It would seem in this case, therefore, that physical force (light) is changed into nascent chemical force, and this nascent chemical force, under the peculiar conditions present, forms organic matter, and reappears as vital force. Light falling on living green leaves is destroyed or consumed in doing the work of decomposition; disappears as light, to reappear as nascent chemical energy; and this in its turn disappears in forming organic matter, to reappear as the vital force of the organic matter thus formed. The light which disappears is proportioned to the O, or the O and H rejected; is proportioned also to the quantity of organic matter formed, and also to the amount of vital force resulting. To illustrate: In the case of amyloids, oxygen-excess falling or running down from plane No. 2 to plane No. 1 generates force to raise C, H, and O, from plane No. 2 to plane No. 3. In the case of albuminoids, oxygen-excess and hydrogen-excess running down from No. 2 to No. 1 generate force to raise C, H, O, and N, from No. 2 to No. 3. To illustrate again: As sun-heat falling upon water disappears as heat, to reappear as mechanical power, raising the water into the clouds, so sunlight falling upon green leaves disappears as light, to reappear as vital force lifting matter from the mineral into the organic kingdom.

2. Germination.—Growing plants, it is seen, take their life-force from the sun; but seeds germinate and commence to grow in the dark. Evidently there must be some other source from which they draw their supply of force. They cannot draw force from the sun. This fact is intimately connected with another fact, viz., that they do not draw their food from the mineral kingdom. The seed in germination feeds entirely upon a

supply of organic matter laid up for it by the mother-plant. It is the decomposition of this organic matter which supplies the force of germination. Chemical compounds are comparatively stable—it requires sun-light to tear them asunder; but organic matter is more easily decomposed—it is almost spontaneously decomposed. It may be that heat (a necessary condition of germination) is the force which determines the decomposition. However this may be, it is certain that a portion of the organic matter laid up in the seed is decomposed, burned up, to form CO_2 and H_2O, and that this combustion furnishes the force by which the mason-work of tissue-making is accomplished. In other words, of the food laid up in the form of starch, dextrine, protoplasm, a portion is decomposed to furnish the force by which the remainder is organized. Hence the seed always loses weight in germination; it cannot develop unless it is in part consumed; "it is not quickened except it die." This self-consumption continues until the leaves and roots are formed; then it begins to draw force from the sun, and food from the mineral kingdom.

To illustrate: In germination, matter running down from plane No. 3 to plane No. 2 generates force by which other similar matter is moved about and raised to a somewhat higher position on plane No. 3. As water raised by the sun may be stored in reservoirs, and in running down from these may do work, so matter raised by sun-force into the organic kingdom by one generation is stored as force to do the work of germination of the next generation. Again, as, in water running through an hydraulic ram, a portion runs to waste, in order to generate force to lift the remainder to a higher level, so, of organic matter stored in the seed, a portion runs to waste to create force to organize the remainder.

Thus, then, it will be seen that three things, viz., the absence of sunlight, the use of organic food, and the loss of weight, are indissolubly connected in germination, and all explained by the principle of conservation of force.

3. Starting of Buds.—Deciduous trees are entirely destitute of leaves during the winter. The buds must start to grow in the spring without leaves, and therefore without drawing force from the sun. Hence, also, food in the organic form must be, and is, laid up from the previous year in the body of the tree. A portion of this is consumed with the formation of CO_2 and H_2O, in order to create force for the development of the buds. So soon as by this means the leaves are formed, the plant begins to draw force from the sun, and food from the mineral kingdom.

4. Pale Plants.—Fungi and etiolated plants have no chlorophyl, therefore cannot draw their force from the sun, nor make organic matters from inorganic.

Hence these also must feed on organic matter; not, indeed, on starch, dextrine, and protoplasm, but on decaying organic matter. In these plants the organic matter is taken up in some form intermediate between the planes No. 3 and No. 2. The matter thus taken up is, a portion of it, consumed with the formation of CO_2 and H_2O, in order to create force necessary to organize the remainder. To illustrate: Matter falling from some intermediate point between No. 2 and No. 3 to No. 2, produces force sufficient to raise matter from the same intermediate point to No. 3; a portion runs to waste downward, and creates force to push the remainder upward.

5. Growth of Green Plants at Night.—It is well known that almost all plants grow at night as well as in the day. It is also known that plants at night exhale CO_2. These two facts have not, however, as far as I know, been connected with one another, and with the principle of conservation of force. It is usually supposed that in the night the decomposition of CO_2 and exhalation of oxygen are checked by withdrawal of sunlight, and some of the CO_2 in the ascending sap is exhaled by a physical law. But this does not account for the growth. It is evident that, in the absence of sunlight, the force required for the work of tissue-building can be derived only from the decomposition and combustion of organic matter. There are two views as to

the source of this organic matter, either or both of which may be correct: First. There seems to be no doubt that most plants, especially those grown in soils rich in *humus*, take up a portion of their food in the form of semi-organic matter, or soluble *humus*. The combustion of a portion of this in every part of the plant, by means of oxygen also absorbed by the roots, and the formation of CO_2, undoubtedly creates a supply of force night and day, independently of sunlight. The force thus produced by the combustion of a portion might be used to raise the remainder into starch, dextrine, etc., or might be used in tissue-building. During the day, the CO_2 thus produced would be again decomposed in the leaves by sunlight, and thus create an additional supply of force. During the night, the CO_2 would be exhaled.*

Again: It is possible that more organic matter is made by sunlight during the day than is used up in tissue-building. Some of this excess is again consumed, and forms CO_2 and H_2O, in order to continue the tissue-building process during the night. Thus the plant during the day stores up sun-force sufficient to do its work during the night. It has been suggested by Dr. J. C. Draper,† though not proved, or even rendered probable,

* For more full account, see my paper, *American Journal of Science*, November, 1859, sixth and seventh heads.

† *American Journal of Science*, November, 1872. The experiments of Dr. Draper are inconclusive, because they are made on *seedlings*, which,

that the force of tissue-building (*force plastique*) is always derived from decomposition, or combustion of organic matter. In that case, the force of organic-matter formation is derived from the sun, while the force of tissue-building (which is relatively small) is derived from the combustion of organic matter thus previously formed.

6. FERMENTATION.—The plastic matters out of which vegetable tissue is built, and which are formed by sunlight in the leaves, are of two kinds, viz., amyloids (dextrine, sugar, starch, cellulose), and albuminoids, or protoplasm. Now, the amyloids are comparatively stable, and do not spontaneously decompose; but the albuminoids not only decompose spontaneously themselves, but drag down the amyloids with which they are associated into concurrent decomposition—not only change themselves, but propagate a change into amyloids. Albuminoids, in various stages and kinds of decomposition, are called ferments. The propagated change in amyloids is called fermentation. By various kinds of ferments, amyloids are thus dragged down step by step to the mineral kingdom, viz., to CO_2 and H_2O. The accompanying table exhibits the various stages of the descent of starch, and the ferments by which they are effected:

until their supply of organic food is exhausted, are independent of sunlight.

1. Starch...	Diastase.
2. Dextrine................	
3. Sugar...................	
4. Alcohol and CO_2.........	Yeast.
5. Acetic acid..............	Mother of vinegar.
6. CO_2 and H_2O............	Mould.

By appropriate means, the process of descent may be stopped on any one of these planes. By far too much is, unfortunately, stopped on the fourth plane. The manufacturer and chemist may determine the downward change through all the planes, and the chemist has recently succeeded in ascending again to No. 4; but the plant ascends and descends the scale at pleasure (avoiding, however, the fourth and fifth), and even passes at one step from the lowest to the highest.

Now, it will be seen by the table that, connected with each of these descensive changes, there is a peculiar ferment associated. Diastase determines the change from starch to dextrine and sugar — saccharification; yeast, the change from sugar to alcohol—fermentation; mother of vinegar, the change from alcohol to acetic acid—acetification; and a peculiar mould, the change from acetic acid to CO_2 and water. But what is far more wonderful and significant is, that, associated with each of these ferments, except diastase, and therefore with each of these descensive changes, except the change from starch to sugar, or saccharification, there is a peculiar form of life. Associated with alcoholic fermentation, there is the yeast-plant; with acetification, the vinegar-plant; and with the decomposition of vinegar, a peculiar kind of mould. We will take the one which is best understood, viz., yeast-plant (saccharomyce), and its relation to alcoholic fermentation.

It is well known that, in connection with alcoholic fermentation, there is a peculiar unicelled plant which grows and multiplies. Fermentation never takes place without the presence of this plant; this plant never grows without producing fermentation, and the rapidity of the fermentation is in exact proportion to the rapidity of the growth of the plant. But, as far as I know, the fact has not been distinctly brought out that the decomposition of the sugar into alcohol and carbonic acid furnishes the force by which the plant grows and multiplies. If the growing cells of the yeast-plant be observed under the microscope, it will be seen that the carbonic-acid bubbles form, and therefore probably the decomposition of sugar takes place only in contact with the surface of the yeast-cells. The yeast-plant not only assimilates matter, but also force. It decomposes the sugar in order that it may assimilate the chemical force set free.

We have already said that the change from starch to sugar, determined by diastase (saccharification), is the only one in connection with which there is no life.

Now, it is a most significant fact, in this connection, that this is also the only change which is not, in a proper sense, descensive, or, at least, where there is no decomposition.

We now pass from the phenomena of vegetable to the phenomena of animal life.

7. Development of the Egg in Incubation.—The development of the egg in incubation is very similar to the germination of a seed. An egg consists of albuminous and fatty matters, so inclosed that, while oxygen of the air is admitted, nutrient matters are excluded. During incubation the egg changes into an embryo; it passes from an almost unorganized to a highly-organized condition, from a lower to a higher condition. There is work done: there must be expenditure of force; but, as we have already seen, vital force is always derived from decomposition. But, as the matters to be decomposed are not taken *ab extra*, the egg must consume itself; that it does so, is proved by the fact that in incubation the egg absorbs oxygen, eliminates CO_2 and probably H_2O, and loses weight. As in the seed, a portion of the matters contained in the egg is consumed in order to create force to organize the remainder. Matter runs down from plane No. 4 to plane No. 2, and generates force to do the work of organization on plane No. 4. The amount of CO_2 and H_2O formed, and therefore the loss of weight, is a measure of the amount of plastic work done.

8. Development within the Chrysalis Shell.—It is well known that many insects emerge from the egg not in their final form, but in a wormlike form, called a larva. After this they pass into a second passive state, in which they are again covered with a kind of shell—a sort of second egg-state, called the chrysalis. From this they again emerge as the perfect insect. The butterfly is the most familiar, as well as the best, illustration of these changes. The larva or caterpillar eats with enormous voracity, and grows very rapidly. When its growth is complete, it covers itself with a shell, and remains perfectly passive and almost immovable for many days or weeks. During this period of quiescence of animal functions there are, however, the most important changes going on within. The wings and legs are formed, the muscles are aggregated in bundles for moving these appendages, the nervous system is more highly developed, the mouth-organs and alimentary canal are greatly changed and more highly organized, the simple eyes are changed into compound eyes. Now, all this requires expenditure of force, and therefore decomposition of matter; but no food is taken, therefore the chrysalis must consume its own substance, and therefore lose weight. It does so; the weight of the emerging butterfly is in many cases not one-tenth that of the caterpillar. Force is stored up in the form of organic matter only to be consumed in doing plastic work.

9. Mature Animals.—Whence do animals derive their vital force? I answer, from the decomposition of their food and the decomposition of their tissues.

Plants, as we have seen, derive their vital force from the decomposition of their mineral food. But the chemical compounds on which plants feed are very stable. Their decomposition requires a peculiar and complex contrivance for the reception and utilization of sunlight. These conditions are wanting in animals. Animals, therefore, cannot feed on chemical compounds of the mineral kingdom; they must have organic food which easily runs into decomposition; they must feed on the vegetable kingdom.

Animals are distinguished from vegetables by incessant decay in every tissue—a decay which is proportional to animal activity. This incessant decay necessitates incessant repair, so that the animal body has been likened to a temple on which two opposite forces are at work in every part, the one tearing down, the other repairing the breach as fast as made. In vegetables no such incessant decay has ever been made out. If it exists, it must be very trifling in comparison. Protoplasm, it is true, is taken up from the older parts of vegetables, and these parts die; but the protoplasm does not seem to decompose, but is used again for tissue-building. Thus the internal activity of animals is of two kinds, tissue-destroying and tissue-building; while that of plants seems to be, principally, at least, of one kind, tissue-building. Animals use food for force and repair and growth, and in the mature animal only for force and repair. Plants, except in reproduction, use food almost wholly for growth—they never stop growing.

Now, the food of animals is of two kinds, amyloids and albuminoids. The carnivora feed entirely on albuminoids; herbivora on both amyloids and albuminoids. All this food comes from the vegetable kingdom, directly in the case of herbivora, indirectly in the case of carnivora. Animals cannot make organic matter. Now, the tissues of animals are wholly albuminoid. It is obvious, therefore, that for the repair of the tissues the food must be albuminoid. The amyloid food, therefore (and, as we shall see in carnivora, much of the albuminoid), must be used wholly for force. As coal or wood, burned in a steam-engine, changes chemical into mechanical energy, so food, in excess of what is used for repair, is burned up to produce animal activity. Let us trace more accurately the origin of animal force by examples.

10. Carnivora.—The food of carnivora is entirely albuminoid. The idea of the older physiologists, in regard to the use of this food, seems to have been as follows: Albuminoid matter is exceedingly unstable; it is matter raised, with much difficulty and against chemical forces, high, and delicately balanced on a pinnacle, in a

state of unstable equilibrium, for a brief time, and then rushes down again into the mineral kingdom. The animal tissues, being formed of albuminoid matter, are short-lived; the parts are constantly dying and decomposing; the law of death necessitates the law of reproduction; decomposition necessitates repair, and therefore food for repair. But the force by which repair is effected was for them, and for many physiologists now, underived, innate. But the doctrine maintained by me in the paper referred to is, that the decomposition of the tissues creates not only the necessity, but also the force, of repair.

Suppose, in the first place, a carnivorous animal uses just enough food to repair the tissues, and no more—say an ounce. Then I say the ounce of tissue decayed not only necessitates the ounce of albuminous food for repair, but the decomposition sets free the force by which the repair is effected. But it will be perhaps objected that the force would all be consumed in repair, and none left for animal activity of all kinds. I answer: it would not all be used up in repair, for, the food being already albuminoid, there is probably little expenditure of force necessary to change it into tissue; while, on the other hand, the force generated by the decomposition of tissue into CO_2, H_2O, and urea, is very great—the ascensive change is small, the descensive change is great. The decomposition of one ounce of albuminous tissue into

CO_2, H_2O, and urea, would therefore create force sufficient not only to change one ounce of albuminous matter into tissue, but also leave a considerable amount for animal activities of all kinds. A certain quantity of matter, running down from plane No. 4 to plane No. 2, creates force enough not only to move the same quantity of matter about on plane No. 4, but also to do much other work besides. It is probable, however, that the wants of animal activity are so immediate and urgent that, under these conditions, much food would be burned for this purpose, and would not reach the tissues, and the tissues would be imperfectly repaired, and would therefore waste.

Take, next, the carnivorous animal full fed. In this case there can be no doubt that, while a portion of the food goes to repair the tissues, by far the larger portion is consumed in the blood, and passes away partly as CO_2 and H_2O through the lungs, and partly as urea through the kidneys. This part is used, and can be of use only, to create force. The food of carnivora, therefore, goes partly to tissue-building, and partly to create heat and force. The force of carnivorous animals is derived partly from decomposing tissues and partly from food-excess consumed in the blood.

11. HERBIVORA.—The food of herbivora and of man is mixed—partly albuminoid and partly amyloid. In man, doubtless, the albuminoids are usually in excess of

what is required for tissue-building; but in herbivora, probably, the albuminoids are not in excess of the requirements of the decomposing tissues. In this case, therefore, the whole of the albuminoids is used for tissue-making, and the whole of the amyloids for force-making. In this class, therefore, these two classes of food may be called tissue-food and force-food. The force of these animals, therefore, is derived partly from the decomposition of the tissues, but principally from the decomposition and combustion of the amyloids and fats.

Some physiologists speak of the amyloid and fat food as being burned to keep up the animal heat; but it is evident that the prime object in the body, as in the steam-engine, is not heat, but force. Heat is a mere condition and perhaps a necessary concomitant of the change, but evidently not the prime object. In tropical regions the heat is not wanted. In the steam-engine, chemical energy is first changed into heat, and heat into mechanical energy; in the body the change is, probably, much of it direct, and not through the intermediation of heat.

12. We see at once, from the above, why it is that plants cannot feed on elements, viz., because their food must be decomposed in order to create the organic matter out of which all organisms are built. This elevation of matter, which takes place in the green

leaves of plants, is the starting-point of life; upon it alone is based the possibility of the existence of the organic kingdom. The running down of the matter there raised determines the vital phenomena of germination, of pale plants, and even of some of the vital phenomena of green plants, and all the vital phenomena of the animal kingdom. The stability of chemical compounds, usable as plant-food, is such that a peculiar contrivance and peculiar conditions found only in the green leaves of plants are necessary for their decomposition. We see, therefore, also, why animals as well as pale plants cannot feed on mineral matter.

We easily see also why the animal activity of carnivora is greater than that of herbivora, for the amount of force necessary for the assimilation of their albuminoid food is small, and therefore a larger amount is left over for animal activity. Their food is already on plane No. 4; assimilation, therefore, is little more than a *shifting* on the plane No. 4 from a liquid to a solid condition —from liquid albuminoid of the blood to solid albuminoid of the tissues.

We see also why the internal activity of plants may conceivably be only of one kind; for, drawing their force from the sun, tissue-making is not necessarily dependent on tissue-decay. While, on the other hand, the internal activity of animals must be of two kinds, decay and repair; for animals always draw a portion of their force, and starving animals the whole of their force, from decaying tissue.

13. There are several general thoughts suggested by this subject, which I wish to present in conclusion :

a. We have said there are four planes of matter raised one above the other : 1. Elements ; 2. Chemical compounds ; 3. Vegetables ; 4. Animals. Their relative position is truly represented thus :

4. *Animals.*
3. *Plants.*
2. *Chemical compounds.*
1. *Elements.*

Now, there are also four planes of force similarly related to each other, viz., physical force, chemical force, vitality, and will. On the first plane of matter operates physical force only ; for chemical force immediately raises matter into the second plane. On the second plane operates, in addition to physical, also chemical force. On the third plane operates, in addition to physical and chemical, also vital force. On the fourth plane, in addition to physical, chemical, and vital, also the force characteristic of animals, viz., will.* With each eleva-

* I might add still another plane and another force, viz., the human plane, on which operate, in addition to all the lower forces, also free-will and reason. I do not speak of these, only because they lie beyond the present ken of inductive science.

tion there is a peculiar force added to the already existing, and a peculiar group of phenomena is the result. As matter only rises step by step from plane to plane, and never two steps at a time, so also force, in its transformation into higher forms of force, rises only step by step. Physical force does not become vital except through chemical force, and chemical force does not become will except through vital force.

Again, we have compared the various grades of matter, not to a gradually rising inclined plane, but to successive planes raised one above the other. There are, no doubt, some intermediate conditions; but, as a broad, general fact, the changes from plane to plane are sudden. Now, the same is true also of the forces operating on these planes—of the different grades of force, and their corresponding groups of phenomena. The change from one grade to another, as from physical to chemical, or from chemical to vital, is not, as far as we can see, by sliding scale, but suddenly. The groups of phenomena which we call physical, chemical, vital, animal, rational, and moral, do not merge into each other by insensible gradations. In the ascensive scale of forces, in the evolution of the higher forces from the lower, there are places of rapid, paroxysmal change.

b. Vital force is transformed into physical and chemical forces; but it is not on that account identical with physical and chemical force, and therefore we ought not, as some would have us, discard the term vital force. There are two opposite errors on this subject: one is the old error of regarding vital force as something innate, underived, having no relation to the other forces of Nature; the other is the new error of regarding the forces of the living body as nothing but ordinary physical and chemical forces, and therefore insisting that the use of the term vital force is absurd and injurious to science. The old error is still prevalent in the popular mind, and still haunts the minds of many physiologists; the new error is apparently a revulsion from the other, and is therefore common among the most advanced scientific minds. There are many of the best scientists who ridicule the use of the term vital force, or vitality, as a remnant of superstition; and yet the same men use the words gravity, magnetic force, chemical force, physical force, etc. Vital force is not underived—is not unrelated to other forces—is, in fact, correlated with them; but it is nevertheless a distinct form of force, far more distinct than any other form, unless it be still higher forms, and therefore better entitled to a distinct name than any lower form. Each form of force gives rise to a peculiar group of phenomena, and the study of these to a peculiar department of science. Now, the group of phenomena called vital is more peculiar, and more different from other groups, than these are from each other; and the science of physiology is a more distinct department than either physics or chemistry; and therefore the form of force which determines these phenomena is more distinct, and better entitled to a distinct name, than either physical or chemical forces. De Candolle, in a recent paper,* suggests the term vital movement instead of vital force; but can we conceive of movement without force? And, if the movement is peculiar, so also is the form of force.

c. Vital is transformed physical and chemical forces; true, but the necessary and very peculiar condition of this transformation is the previous existence then and there of living matter. There is something so wonderful in this peculiarity of vital force that I must dwell on it a little.

Elements brought in contact with each other under certain physical conditions—perhaps heat or electricity—unite and rise into the second plane, i. e., of chemical compounds; so also several elements, C, H, O, and N, etc., brought in contact with each other under certain physical or chemical conditions, such as light, nascency, etc., unite and rise into plane No. 3, i. e., form organic matter. In both cases there is chemical union under certain physical conditions; but in the latter there is one unique condition, viz., the previous existence then and there of organic matter, under the guidance of which the transformation of matter takes place. In a

* *Archives des Sciences*, vol. xlv., p. 345, December, 1872.

word, organic matter is necessary to produce organic matter; there is here a law of like producing like—there is an assimilation of matter.

Again, physical force changes into other forms of physical force, or into chemical force, under certain physical conditions; so also physical and chemical forces are changed into vital force under certain physical conditions. But, in addition, there is one altogether unique condition of the latter change, viz., the previous existence then and there of vital force. Here, again, like produces like—here, again, there is assimilation of force.

This law of like producing like—this law of assimilation of matter and force—runs throughout all vital phenomena, even to the minutest details. It is a universal law of generation, and determines the existence of species; it is the law of formation of organic matter and organic force; it determines all the varieties of organic matter which we call tissues and organs, and all the varieties of organic force which we call functions. The same nutrient pabulum, endowed with the same properties and powers, carried to all parts of a complex organism by this wonderful law of like producing like, is changed into the most various forms and endowed with the most various powers. There are certainly limits and exceptions to this law, however; otherwise differentiation of tissues, organs, and functions, could not take place in embryonic development; but the limits and exceptions are themselves subject to a law even more wonderful than the law of like producing like itself, viz., the law of evolution. There is in all organic nature, whether organic kingdom, organic individual, or organic tissues, a law of variation, strongest in the early stages, limited very strictly by another law —the law of inheritance, of like producing like.

d. We have seen that all development takes place at the expense of decay—all elevation of one thing, in one place, at the expense of corresponding running down of something else in another place. Force is only transferred and transformed. The plant draws its force from the sun, and therefore what the plant gains the sun loses. Animals draw from plants, and therefore what the animal kingdom gains the vegetable kingdom loses. Again, an egg, a seed, or a chrysalis, developing to a higher condition, and yet taking nothing *ab extra*, must lose weight. Some part must run down, in order that the remainder should be raised to a higher condition. The amount of evolution is measured by the loss of weight. By the law of conservation of force, it is inconceivable that it should be otherwise. Evidently, therefore, in the universe, taken as a whole, evolution of one part must be at the expense of some other part. The evolution or development of the whole cosmos—of the whole universe of matter—as a unit, by forces within itself, according to the doctrine of conservation of force, is inconceivable. If there be any such evolution, at all comparable with any known form of evolution, it can only take place by a constant increase of the whole sum of energy, i. e., by a constant influx of divine energy—for the same quantity of matter in a higher condition must embody a greater amount of energy.

e. Finally, as organic matter is so much matter taken from the common fund of matter of earth and air, embodied for a brief space, to be again by death and decomposition returned to that common fund, so also it would seem that the organic forces of the living bodies of plants and animals may be regarded as so much force drawn from the common fund of physical and chemical forces, to be again all refunded by death and decomposition. Yes, by decomposition; we can understand this. But death! can we detect any thing returned by simple death? What is the nature of the difference between the living organism and a dead organism? We can detect none, physical or chemical. All the physical and chemical forces withdrawn from the common fund of Nature, and embodied in the living organism, seem to be still embodied in the dead until little by little it is returned by decomposition. Yet the difference is immense, is inconceivably great. What is the nature of this difference expressed in the formula of material science? What is it that is gone, and whither is it gone?

There is something here which science cannot yet understand. Yet it is just this loss which takes place in death, and before decomposition, which is in the highest sense vital force.

Let no one from the above views, or from similar views expressed by others, draw hasty conclusions in favor of a pure materialism. Force and matter, or spirit and matter, or God and Nature, these are the opposite poles of philosophy—they are the opposite poles of thought. There is no clear thinking without them. Not only religion and virtue, but science and philosophy, cannot even exist without them. The belief in spirit, like the belief in matter, rests on its own basis of phenomena. The true domain of philosophy is to reconcile these with each other.

CORRELATION OF NERVOUS AND MENTAL FORCES.

By ALEXANDER BAIN, LL. D.,

PROFESSOR OF LOGIC AND MENTAL PHILOSOPHY IN THE UNIVERSITY OF ABERDEEN.

THE CORRELATION OF NERVOUS AND MENTAL FORCES.

The doctrine called the correlation, persistence, equivalence, transmutability, indestructibility of force, or the conservation of energy, is a generality of such compass that no single form of words seems capable of fully expressing it; and different persons may prefer different statements of it. My understanding of the doctrine is, that there are five chief powers or forces in Nature: one *mechanical*, or *molar*, the momentum of moving matter; the others *molecular*, or embodied in the molecules, also supposed in motion—these are, heat, light, chemical force, electricity. To these powers, which are unquestionable and distinct, it is usual to add vital force, of which, however, it is difficult to speak as a whole; but one member of our vital energies, the nerve-force, allied to electricity, fully deserves to rank in the correlation.

Taking the one mechanical force, and those three of

the molecular named heat, chemical force, electricity, there has now been established a definite rate of commutation, or exchange, when any one passes into any other. The mechanical equivalent of heat, the 772 foot-pounds of Joule, expresses the rate of exchange between mechanical momentum and heat: the equivalent or exchange of heat and chemical force is given (through the researches of Andrews and others) in the figures expressing the heat of combinations; for example, one pound of carbon burnt evolves heat enough to raise 8,080 pounds of water one degree, C. The combination of these to equivalents would show that the consumption of half a pound of carbon would raise a man of average weight to the highest summit of the Himalayas.

It is an essential part of the doctrine, that force is never absolutely created, and never absolutely destroyed, but merely transmuted in form or manifestation.

As applied to living bodies, the following are the usual positions. In the growth of plants, the forces of the solar ray—heat and light—are expended in decomposing (or deoxidizing) carbonic acid and water, and in building up the living tissues from the liberated carbon and the other elements; all which force is given up when these tissues are consumed, either as fuel in ordinary combustion, or as food in animal combustion.

It is this animal combustion of the matter of plants, and of animals (fed on plants)—namely, the reoxidation of carbon, hydrogen, etc.—that yields all the manifestations of power in the animal frame. And, in particular, it maintains (1) a certain warmth or temperature of the whole mass, against the cooling power of surrounding space; it maintains (2) mechanical energy, as muscular power; and it maintains (3) nervous power, or a certain flow of the influence circulating through the nerves, which circulation of influence, besides reacting on the other animal processes—muscular, glandular, etc.—has for its distinguishing concomitant the MIND.

The extension of the correlation of force to mind, if at all competent, must be made through the nerve-force, a genuine member of the correlated group. Very serious difficulties beset the proposal, but they are not insuperable.

The history of the doctrines relating to mind, as connected with body, is in the highest degree curious and instructive, but, for the purpose of the present paper, we shall notice only certain leading stages of the speculation.*

Not the least important position is the Aristotelian; a position in some respects sounder than what followed and grew out of it. In Aristotle, we have a kind of gradation from the life of plants to the highest form of

* For the fuller elaboration of the point here referred to, see Chapter VII., Professor Bain's "Mind and Body"—an earlier volume in the present series.

human intelligence. In the following diagram, the continuous lines may represent the material substance, and the dotted lines the immaterial:

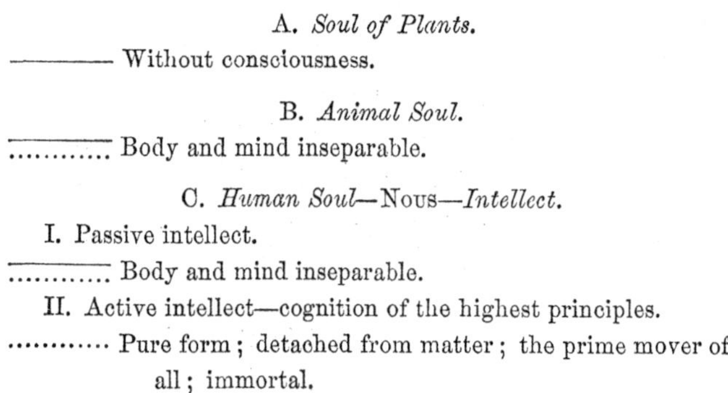

All the phases of life and mind are inseparably interwoven with the body (which inseparability is Aristotle's definition of the soul) except the last, the active *nous*, or intellect, which is detached from corporeal matter, self-subsisting, the essence of Deity, and an immortal substance, although the immortality is not personal to the individual. (The immateriality of this higher intellectual agent was net, however, that thorough-going negation of all material attributes which we now understand by the word "immaterial.") How such a self-subsisting and purely spiritual soul could hold communication with the body-leagued souls, Aristotle was at a loss to say—the difficulty reappeared after him, and has never been got over. That there should be an agency totally apart from, and entirely transcending, any known powers of inert matter, involves no difficulty—for who is to limit the possibilities of existence? The perplexity arises only when this radically new and superior principle is made to be, as it were, off and on with the material principle; performing some of its functions in pure isolation, and others of an analogous kind by the aid of the lower principle. The difference between the active and the passive reason of Aristotle is a mere difference of gradation; the supporting agencies assumed by him are a total contrast in kind—wide as the poles asunder. There is no breach of continuity in the phenomena, there is an impassable chasm between their respective foundations.

Fifteen centuries after Aristotle, we reach what may be called the modern settlement of the relations of mind and body, effected by Thomas Aquinas. He extended the domain of the independent immaterial principle from the highest intellectual soul of Aristotle to all the three souls recognized by him—the vegetable or plant soul (without consciousness), the animal soul (with consciousness), and the intellect throughout. The two lower souls—the vegetable and the animal—need the coöperation of the body in this life; the intellect works without any bodily organ, except that it makes use of the perceptions of the senses.

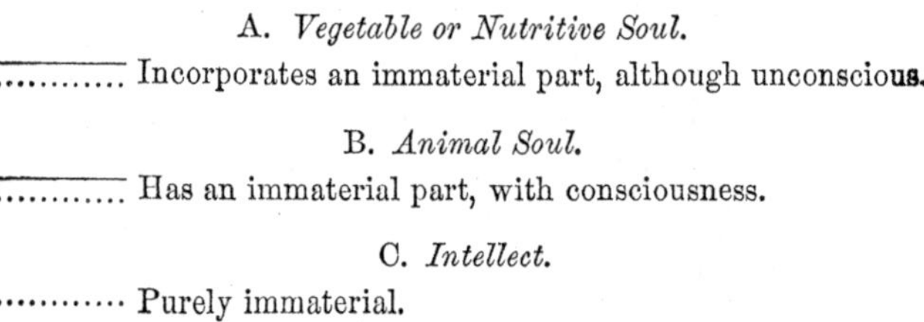

The animal soul, B, contains sensation, appetite, and emotion, and is a mixed or two-sided entity; but the intellect, C, is a purely one-sided entity, the immaterial. This does not relieve our perplexities; the phenomena are still generically allied and continuous — sensation passes into intellect without any breach of continuity; but as regards the agencies, the transition from a mixed or united material and immaterial substance to an immaterial substance apart, is a transition to a differently constituted world, to a transcendental sphere of existence.

The settlement of Aquinas governed all the schools and all the religious creeds, until quite recent times; it is, for example, substantially the view of Bishop Butler. At the instance of modern physiology, however, it has undergone modifications. The dependence of purely intellectual operations, as memory, upon the material processes, has been reluctantly admitted by the partisans of an immaterial principle; an admission incompatible with the isolation of the intellect in Aristotle and in Aquinas. This more thorough-going connection of the mental and the physical has led to a new form of expressing the relationship, which is nearer the truth, without being, in my judgment, quite accurate. It is now often said *the mind and the body act upon each other;* that neither is allowed, so to speak, to pursue its course alone—there is a constant interference, a mutual influence between the two. This view is liable to the following objections:

1. In the first place, it assumes that we are entitled to speak of mind apart from body, and to affirm its powers and properties in that separate capacity. But of mind apart from body we have no direct experience, and absolutely no knowledge. The wind may act upon the sea, and the waves may react upon the wind; but the agents are known in separation—they are seen to exist apart before the shock of collision; but we are not permitted to see a mind acting apart from its material companion.

2. In the second place, we have every reason for believing that there is an unbroken material succession, side by side with all our mental processes. From the ingress of a sensation, to the outgoing responses in action, the mental succession is not for an instant dissevered from a physical succession. A new prospect bursts upon the view; there is a mental result of sensations, emotion, thought, terminating in outward displays of speech or gesture. Parallel to this mental

series is the physical series of facts, the successive agitation of the physical organs, called the eye, the retina, the optic nerve, optic centres, cerebral hemispheres, outgoing nerves, muscles, etc. There is an unbroken physical circle of effects, maintained while we go the round of the mental circle of sensation, emotion, and thought. It would be incompatible with every thing we know of the cerebral action to suppose that the physical chain ends abruptly in a physical void, occupied by an immaterial substance; which immaterial substance, after working alone, imparts its results to the other edge of the physical break, and determines the active response —two shores of the material with an intervening ocean of the immaterial. There is, in fact, no rupture of nervous continuity. The only tenable supposition is, that mental and physical proceed together, as individual twins. When, therefore, we speak of a mental cause, a mental agency, we have always a two-sided cause; the effect produced is not the effect of mind alone, but of mind in company with body. That mind should have operated on the body, is as much as to say that a two-sided phenomenon, one side being bodily, can influence the body; it is, after all, body acting upon body. When a shock of fear paralyzes digestion, it is not the emotion of fear, in the abstract, or as a pure mental existence, that does the harm; it is the emotion in company with a peculiarly excited condition of the brain and nervous system; and it is this condition of the brain that deranges the stomach. When physical nourishment, or physical stimulant, acting through the blood, quiets the mental irritation, and restores a cheerful tone, it is not a bodily fact causing a mental fact by a direct line of causation: the nourishment and the stimulus determine the circulation of blood to the brain, give a new direction to the nerve-currents, and the mental condition corresponding to this particular mode of cerebral action henceforth manifests itself. The line of mental sequence is thus, not mind causing body, and body causing mind, but mind-body giving birth to mind-body; a much more intelligible position. For this double or conjoint causation, we can produce evidence; for the single-handed causation we have no evidence.

If it were not my peculiar province to endeavor to clear up the specially metaphysical difficulties of the relationship of mind and body, I would pass over what is to me the most puzzling circumstance of the relationship, and indeed the only real difficulty in the question.

I say the real difficulty, for factitious difficulties in abundance have been made out of the subject. It is made a mystery how mental functions and bodily functions should be allied together at all. That, however, is no business of ours; we accept this alliance, as we do any other alliance, such as gravity with inert matter, or light with heat. As a fact of the universe, the union is, properly speaking, just as acceptable, and as intelligible, as the separation would be, if that were the fact. The real difficulty is quite another thing.

What I have in view is this: when I speak of mind as allied with body—with a brain and its nerve-currents—I can scarcely avoid *localizing* the mind, giving it a local habitation. I am thereupon asked to explain what always puzzled the schoolmen, namely, whether the mind is all in every part, or only all in the whole; whether in tapping any point I may come at consciousness, or whether the whole mechanism is wanted for the smallest portion of consciousness. One might perhaps turn the question by the analogy of the telegraph wire, or the electric circuit, and say that a complete circle of action is necessary to any mental manifestation; which is probably true. But this does not meet the case. The fact is that, all this time we are speaking of nerves and wires, we are not speaking of mind, properly so called, at all; we are putting forward physical facts that go along with it, but these physical facts are not the mental fact, and they even preclude us from thinking of the mental fact. We are in this fix: mental states and bodily states are utterly contrasted; they cannot be compared, they have nothing in common except the most general of all attributes, degree, and order in time; when engaged with one we must be oblivious of all that distinguishes the other. When I am studying a brain and nerve communicating, I am engrossed with properties exclusively belonging to the object or material world; I am at that moment (except by very rapid transitions or alternations) unable to conceive a truly mental fact, my truly mental consciousness. Our mental experience, our feelings and thoughts, have no extension, no place, no form or outline, no mechanical division of parts; and we are incapable of attending to any thing mental until we shut off the view of all that. Walking in the country in spring, our mind is occupied with the foliage, the bloom, and the grassy meads, all purely objective things; we are suddenly and strongly arrested by the odor of the May-blossom; we give way for a moment to the sensation of sweetness: for that moment the objective regards cease; we think of nothing extended; we are in a state where extension has no footing; there is, to us, place no longer. Such states are of short duration, mere fits, glimpses; they are constantly shifted and alternated with object states, but while they last and have their full power we are in a different world; the material world is blotted out, eclipsed, for the instant unthinkable. These subject-moments are studied to advantage in bursts of intense pleasure, or intense pain, in fits of engrossed reflection, especially reflection upon mental facts; but they are seldom sustained in purity beyond a very short interval; we are constantly returning to the object-side of things—

to the world where extension and place have their being.

This, then, as it appears to me, is the only real difficulty of the physical and mental relationship. There is an alliance with matter, with the object, or extended world; but the thing allied, the mind proper, has itself no extension, and cannot be joined in local union. Now, we have no form of language, no familiar analogy, suited to this unique conjunction; in comparison with all ordinary unions, it is a paradox or a contradiction. We understand union in the sense of local connection; here is a union where local connection is irrelevant, unsuitable, contradictory, for we cannot think of mind without putting ourselves out of the world of place. When, as in pure feeling—pleasure or pain—we change to the subject attitude from the object attitude, we have undergone a change not to be expressed by place; the fact is not properly described by the transition from the *external* to the *internal*, for that is still a change in the region of the extended. The only adequate expression is a *change of state:* a change from the state of the extended cognition to a state of unextended cognition. By various theologians, heaven has been spoken of as not a place, but a *state;* and this is the only phrase that I can find suitable to describe the vast, though familiar and easy, transition from the material or extended, to the immaterial or unextended side of the universe of being.

When, therefore, we talk of incorporating mind with brain, we must be held as speaking under an important reserve or qualification. Asserting the union in the strongest manner, we must yet deprive it of the almost invincible association of union in place. An extended organism is the condition of our passing into a state where there is no extension. A human being is an extended and material thing, attached to which is the power of becoming alive to feeling and thought, the extreme remove from all that is material; a condition of *trance* wherein, while it lasts, the material drops out of view—so much so, that we have not the power to represent the two extremes as lying side by side, as container and contained, or in any other mode of local conjunction. The condition of our existing thoroughly in the one, is the momentary eclipse or extinction of the other.

The only mode of union that is not contradictory is the union of close succession in *time;* or of position in a continued thread of conscious life. We are entitled to say that the same being is, by alternate fits, object and subject, under extended and under unextended consciousness; and that without the extended consciousness the unextended would not arise. Without certain peculiar modes of the extended—what we call a cerebral organization, and so on—we could not have those times of trance, our pleasures, our pains, and our ideas, which at present we undergo fitfully and alternately with our extended consciousness.

Having thus called attention to the metaphysical difficulty of assigning the relative position of mind and matter, I will now state briefly what I think the mode of dealing with mind in correlation with the other forces. That there is a definite equivalence between mental manifestations and physical forces, the same as between the physical forces themselves, is, I think, conformable to all the facts, although liable to peculiar difficulties in the way of decisive proof:

I. The mental manifestations are in exact proportion to their physical supports.

If the doctrine of the thorough-going connection of mind and body is good for any thing, it must go this length. There must be a numerically-proportioned rise and fall of the two together. I believe that all the unequivocal facts bear out this proportion.

Take first the more obvious illustrations. In the employment of external agents, as warmth and food, all will admit that the sensation rises exactly as the stimulant rises, until a certain point is reached, when the agency changes its character; too great heat destroying the tissues, and too much food impeding digestion. There is, although we may not have the power to fix it, a *sensational equivalent* of heat, of food, of exercise, of sound, of light; there is a definite change of feeling, an accession of pleasure or of pain, corresponding to a rise of temperature in the air of 10°, 20°, or 30°. And so with regard to every other agent operating upon the human sensibility: there is, in each set of circumstances, a sensational equivalent of alcohol, of odors, of music, of spectacle.

It is this definite relation between outward agents and the human feelings that renders it possible to discuss human interests from the objective side, the only accessible side. We cannot read the feelings of our fellows; we merely presume that like agents will affect them all in nearly the same way. It is thus that we measure men's fortunes and felicity by the numerical amount of certain agents, as money, and by the absence or low degree of certain other agents, the causes of pain and the depressors of vitality. And, although the estimate is somewhat rough, this is not owing to the indefiniteness of the sensational equivalent, but to the complications of the human system, and chiefly to the narrowness of the line that everywhere divides the wholesome from the unwholesome degrees of all stimulants.

Let us next represent the equivalence under vital or physiological action. The chief organ concerned is the brain; of which we know that it is a system of myriads of connecting threads, ramifying, uniting, and crossing at innumerable points; that these threads are actuated or made alive with a current influence called the nerve force; that this nerve-force is a member of the group of correlating forces; that it is immediately derived from the changes in the blood, and in the last resort from oxidation, or combustion, of the materials of the food, of which combustion it is a definite equivalent. We know, further, that there can be no feeling, no volition, no intellect, without a proper supply of blood, containing both oxygen and the material to be oxidized; that, as the blood is richer in quality in regard to these constituents, and more abundant in quantity, the mental processes are more intense, more vivid. We know also that there are means of increasing the circulation in one organ, and drawing it off from another, chiefly by calling the one into greater exercise, as when we exert the muscles or convey food to the stomach; and that, when mental processes are more than usually intensified, the blood is proportionally drawn to the brain; the oxidizing process is there in excess, with corresponding defect and detriment in other organs. In high mental excitement, digestion is stopped; muscular vigor is abated except in the one form of giving vent to the feelings, thoughts, and purposes; the general nutrition languishes; and, if the state were long continued or oft repeated, the physical powers, strictly so called, would rapidly deteriorate. We know, on the other extreme, that sleep is accompanied by reduced circulation in the brain; there is in fact a reduced circulation generally;

while of that reduced amount more goes to the nutritive functions than to the cerebral.

In listening to Dr. Frankland's lecture on "Muscular Power," delivered at the Royal Institution of London, I noticed that, in accounting for the various items of expenditure of the food, he gave "mental work" as one heading, but declined to make an entry thereinunder. I can imagine two reasons for this reserve, the statement of which will further illustrate the general position. In the first place, it might be supposed that mind is a phenomenon so anomalous, uncertain, so remote from the chain of material cause and effect, that it is not even to be mentioned in that connection.

To which I should say, that mind is indeed, as a phenomenon, widely different from the physical forces, but, nevertheless, rises and falls in strict numerical concomitance with these: so that it still enters, if not directly, at least indirectly, into the circle of the correlated forces. Or, secondly, the lecturer may have held that, though a definite amount of the mental manifestations accompanies a definite amount of oxidation in the special organs of mind, there is no means of reducing this to a measure, even in an approximate way. To this I answer, that the thing is difficult but not entirely impracticable. There is a possibility of giving, approximately at least, the amount of blood circulating in the brain, in the ordinary waking state; and, as during a period of intense excitement we know that there is a general reduction, almost to paralysis, of the collective vital functions, we could not be far mistaken in saying that, in that case, perhaps one-half or one-third of all the oxidation of the body was expended in keeping up the cerebral fires.

It is a very serious drawback in any department of knowledge, where there are relations of quantity, to be unable to reduce them to numerical precision. This is the case with mind in a great degree, although not with it alone; many physical qualities are in the same state of unprecise measurement. We cannot reduce to numbers the statement of a man's constitutional vigor, so as to say how much he has lost by fatigue, by disease, by age, or how much he has gained by a certain healthy regimen. Undoubtedly, however, it is in mind that the difficulties of attaining the numerical statement are greatest if not nearly insuperable. When we say that one man is more courageous, more loving, more irascible than another, we apply a scale of degree, existing in our own mind, but so vague that we may apply it differently at different times, while we can hardly communicate it to others exactly as it stands to ourselves. The consequence is, that a great margin of allowance must always be made in those statements; we can never run a close argument, or contend for a nice shade of distinction. Between the extremes of timidity and courage of character the best observer could not entertain above seven or eight varieties of gradation, while two different persons consulting together could hardly agree upon so minute a subdivision as that. The phrenologists, in their scale of qualities, had the advantage of an external indication of size, but they must have felt the uselessness of graduating this beyond the delicacy of discriminating the subjective side of character; and their extreme scale included twenty steps or interpolations.

Making allowance for this inevitable defect, I will endeavor to present a series of illustrations of the principle of correlation as applied to mind, in the manner explained. I deal not with mind directly, but with its material side, with whose activity, measured exactly as we measure the other physical forces, true mental activity has a definite correspondence.

Let us suppose, then, a human being with average physical constitution, in respect of nutritive vigor, and fairly supplied with food and with air, or oxygen. The result of the oxidation of the food is a definite total of force, which may be variously distributed. The demand made by the brain, to sustain the purely mental functions, may be below average, or above average; there will be a corresponding, but inverse, variation of the remainder available for the more strictly physical processes, as muscular power, digestive power, animal heat, and so on.

In the first case supposed, the case of a small demand for mental work and excitement, we look for, and we find, a better *physique*—greater muscular power and endurance, more vigor of digestion, rendering a coarser food sufficient for nourishment, more resistance to excesses of cold and heat; in short, a constitution adapted to physical drudgery and physical hardship.

Take, now, the other extreme. Let there be a great demand for mental work. The oxidation must now be disproportionately expended in the brain; less is given to the muscles, the stomach, the lungs, the skin, and secreting organs generally. There is a reduction of the possible muscular work, and of the ability to subsist on coarser food, and to endure hardship. Experience confirms this inference; the common observation of mankind has recognized the fact—although in a vague, unsteady form—that the head-worker is not equally fitted to be a hand-worker. The master, mistress, or overseer has each more delicacy of sense, more management, more resource, than the manual operatives, but to these belongs the superiority of muscular power and persistence.

There is nothing incompatible with the principle in allowing the possibility of combining, under certain favorable conditions, both physical and mental exertion in considerable amount. In fact, the principle teaches us exactly how the thing may be done. Improve the quality and increase the quantity of the food; increase the supply of oxygen by healthy residence; let the habitual muscular exertion be such as to strengthen and not impair the functions; abate as much as possible all excesses and irregularities, bodily and mental; add the enormous economy of an educated disposal of the forces; and you will develop a higher being, a *greater aggregate* of power. You will then have more to spare for all kinds of expenditure—for the physico-mental, as well as for the strictly physical. What other explanation is needed of the military superiority of the officer over the common soldier? of the general efficiency of the man nourished, but not enervated, by worldly abundance?

It may be possible, at some future stage of scientific inquiry, to compute the comparative amount of oxidation in the brain during severe mental labor. Even now, from obvious facts, we must pronounce it to be a very considerable fraction of the entire work done in the system. The privation of the other interests during mental exertion is so apparent, so extensive, that if the exertion should happen to be long continued, a liberal atonement has to be made in order to stave off general insolvency. Mental excess counts as largely as muscular excess in the diversion of power; it would be competent to suppose either the one or the other reducing the remaining forces of the system to one-half of their proper amount. In both cases, the work of restoration must be on the same simple plan of redressing the inequality, of allowing more than the average flow of blood to the impoverished organs, for a length of time corresponding to the period when their nourishment has been too small. It is in this consideration that we seem to have the reasonable, I may say the arithmetical, basis of the constitutional treatment of chronic disease. We *repay the debt to Nature* by allowing the weakened organ to be better nourished and less taxed, according to the degradation it has undergone by the opposite line of treatment. In a large class of diseases we have obviously a species of insolvency, to be dealt with according to the sound method of readjusting the relations of expenditure and income. And, if such be the true theory, it seems to follow that medication is only an inferior adjunct. Drugs, even in their happiest application, can but guide and favor the restorative process; just as the stirring of a fire may make it burn, provided there be the needful fuel.

There is thus a definite, although not numerically-statable relation, between the total of the physico-mental forces and the total of the purely physical processes. The grand aggregate of the oxidation of the system includes both; and, the more the force taken up by one, the less is left to the other. Such is the statement of the correlation of mind to the other forces of Nature. We do not deal with pure mind—mind in the abstract; we have no experience of an entity of that description. We deal with a compound or two-sided phenomenon—mental on one side, physical on the other; there is a definite correspondence in degree, although a difference of nature, between the two sides; and the physical side is itself in full correlation with the recognized physical forces of the world.

II. There remains another application of the doctrine, perhaps equally interesting to contemplate, and more within my special line of study. I mean the correlation of the mental forces among themselves (still viewed in the conjoint arrangement). Just as we assign limits to mind as a whole, by a reference to the grant of physical expenditure, in oxidation, etc., for the department, so we must assign limits to the different phases or modes of mental work—thought, feeling, and so on—according to the share allotted to each; so that, while the mind as a whole may be stinted by the demands of the non-mental functions, each separate manifestation is bounded by the requirements of the others. This is an inevitable consequence of the general principle, and equally receives the confirmation of experience. There is the same absence of numerical precision of estimate; our scale of quantity can have but few divisions between the highest and the lowest degrees, and these not well fixed.

What is required for this application of the principle is, to ascertain the comparative cost, in the physical point of view, of the different functions of the mind.

The great divisions of the mind are—feeling, will, and thought; feeling, seen in our pleasures and pains; will, in our labors to attain the one and avoid the other; thought, in our sensations, ideas, recollections, reasonings, imaginings, and so on. Now, the forces of the mind, with their physical supports, may be evenly or unevenly distributed over the three functions. They may go by preference either to feeling, to action, or to thinking; and, if more is given to one, less must remain to the others, the entire quantity being limited.

First, as to the feelings. Every throb of pleasure costs something to the physical system; and two throbs cost twice as much as one. If we cannot fix a precise equivalent, it is not because the relation is not definite, but from the difficulties of reducing degrees of pleasure to a recognized standard. Of this, however, there can be no reasonable doubt—namely, that a large amount of pleasure supposes a corresponding large expenditure of blood and nerve-tissue, to the stinting, perhaps, of the active energies and the intellectual processes. It is a matter of practical moment to ascertain what pleasures cost least, for there are thrifty and unthrifty modes of spending our brain and heart's blood. Experience probably justifies us in saying that the narcotic stimulants are, in general, a more extravagant expenditure than the stimulation of food, society, and fine art. One of the safest of delights, if not very acute, is the delight of abounding physical vigor; for, from the very supposition, the supply to the brain is not such as to interfere with the general interests of the system. But the theory of pleasure is incomplete without the theory of pain.

As a rule, pain is a more costly experience than pleasure, although sometimes economical as a check to the spendthrift pleasures. Pain is physically accompanied by an excess of blood in the brain, from at least two causes—extreme intensity of nervous action, and conflicting currents, both being sources of waste. The sleeplessness of the pained condition means that the circulation is never allowed to subside from the brain; the irritation maintains energetic currents, which bring the blood copiously to the parts affected.

There is a possibility of excitement, of considerable amount, without either pleasure or pain; the cost here is simply as the excitement: mere surprises may be of this nature. Such excitement has no value, except intellectually; it may detain the thoughts, and impress the memory, but it is not a final end of our being, as pleasure is; and it does not waste power to the extent that pain does. The ideally best condition is a moderate surplus of pleasure—a gentle glow, not rising into brilliancy or intensity, except at considerable intervals (say a small portion of every day), falling down frequently to indifference, but seldom sinking into pain.

Attendant on strong feeling, especially in constitutions young or robust, there is usually a great amount of mere bodily vehemence, as gesticulation, play of countenance, of voice, and so on. This counts as muscular work, and is an addition to the brain-work. Properly speaking, the cerebral currents discharge themselves in movements, and are modified according to the scope given to those movements. Resistance to the movements is liable to increase the conscious activity of the brain, although a continuing resistance may suppress the entire wave.

Next as to the will, or our voluntary labors and pursuits for the great ends of obtaining pleasure and warding off pain. This part of our system is a compound experience of feeling and movement; the properly mental fact being included under feeling—that is, pleasure and pain, present or imagined. When our voluntary endeavors are successful, a distinct throb of pleasure is the result, which counts among our valuable enjoyments: when they fail, a painful and depressing state ensues. The more complicated operations of the will, as in adjusting many opposite interests, bring in the element of conflict, which is always painful and wasting. Two strong stimulants pointing opposite ways, as when a miser has to pay a high fee to the surgeon that saves his eyesight, occasion a fierce struggle and severe draft upon the physical supports of the feelings.

Although the processes of feeling all involve a manifest, and it may be a serious, expenditure of physical power, which of course is lost to the purely physical functions; and although the extreme degrees of pleasure, of pain, or of neutral excitement, must be adverse to the general vigor; yet the presumption is, that we can afford a certain moderate share of all these without too great inroads on the other interests. It is the thinking or intellectual part of us that involves the heaviest item of expenditure in the physico-mental department. Any thing like a great or general cultivation of the powers of thought, or any occupation that severely and continuously brings them into play, will induce such a preponderance of cerebral activity, in oxidation and in nerve-currents, as to disturb the balance of life, and to require special arrangements for redeeming that disturbance. This is fully verified by all we know of the tendency of intellectual application to exhaust the physical powers, and to bring on early decay.

A careful analysis of the operations of the intellect enables us to distinguish the kind of exercises that involve the greatest expenditure, from the extent and the intensity of the cerebral occupation. I can but make a rapid selection of leading points:

First. The mere exercise of the senses, in the way of attention, with a view to watch, to discriminate, to identify, belongs to the intellectual function, and exhausts the powers according as it is long continued, and according to the delicacy of the operation; the meaning of delicacy being that an exaggerated activity of the organ is needed to make the required discernment. To be all day on the *qui vive* for some very slight and barely perceptible indications to the eye or the ear, as in catching an indistinct speaker, is an exhausting labor of attention.

Secondly. The work of acquisition is necessarily a process of great nervous expenditure. Unintentional imitation costs least, because there is no forcing of reluctant attention. But a course of extensive and various acquisitions cannot be maintained without a large supply of blood to cement all the multifarious connections of the nerve-fibres, constituting the physical side of acquisition. An abated support of other mental functions, as well as of the purely physical functions, must accompany a life devoted to mental improvement, whether arts, languages, sciences, moral restraints, or other culture.

Of special acquisitions, languages are the most ap

parently voluminous; but the memory for visible or pictorial aspects, if very high, as in the painter and the picturesque poet, makes a prodigious demand upon the plastic combinations of the brain.

The acquisition of science is severe, rather than multifarious; it glories in comprehending much in little, but that little is made up of painful abstract elements, every one of which, in the last resort, must have at its beck a host of explanatory particulars: so that, after all, the burden lies in the multitude. If science is easy to a select number of minds, it is because there is a large spontaneous determination of force to the cerebral elements that support it; which force is supplied by the limited common fund, and leaves so much the less for other uses.

If we advert to the moral acquisitions and habits in a well-regulated mind, we must admit the need of a large expenditure to build up the fabric. The carefully-poised estimate of good and evil for self, the ever-present sense of the interests of others, and the ready obedience to all the special ordinances that make up the morality of the time, however truly expressed in terms of high and abstract spirituality, have their counterpart in the physical organism; they have used up a large and definite amount of nutriment, and, had they been less developed, there would have been a gain of power to some other department, mental or physical.

Refraining from further detail on this head, I close the illustration by a brief reference to one other aspect of mental expenditure, namely, the department of intellectual production, execution, or creativeness, to which in the end our acquired powers are ministerial. Of course, the greater the mere continuance or amount of intellectual labor in business, speculation, fine art, or any thing else, the greater the demand on the *physique*. But amount is not all. There are notorious differences of severity or laboriousness, which, when closely examined, are summed up in one comprehensive statement—namely, the number, the variety, and the conflicting nature of the conditions that have to be fulfilled. By this we explain the difficulty of work, the toil of invention, the harassment of adaptation, the worry of leadership, the responsibility of high office, the severity of a lofty ideal, the distraction of numerous sympathies, the meritoriousness of sound judgment, the arduousness of any great virtue. The physical facts underlying the mental fact are a wide-spread agitation of the cerebral currents, a tumultuous conflict, a consumption of energy.

It is this compliance with numerous and opposing conditions that obtains the most scanty justice in our appreciation of character. The unknown amount of painful suppression that a cautious thinker, a careful writer, or an artist of fine taste, has gone through, represents a great physico-mental expenditure. The regard to evidence is a heavy drag on the wings of speculative daring. The greater the number of interests that a political schemer can throw overboard, the easier his work of construction. The absence of restraints—of severe conditions—in fine art, allows a flush and ebullience, an opulence of production, that is often called the highest genius. The Shakespearean profusion of images would have been reduced to one-half, if not less, by the self-imposed restraints of Pope, Gray, or Tennyson. So, reckless assertion is fuel to eloquence. A man of ordinary fairness of mind would be no match for the wit and epigram of Swift.

And again. The incompatibility of diverse attributes, even in minds of the largest compass (which supposes equally large physical resources), belongs to the same fundamental law. A great mind may be great in many things, because the same kind of power may have numerous applications. The scientific mind of a high order is also the practical mind; it is the essence of reason in every mode of its manifestation—the true philosopher in conduct as well as in knowledge. On such a mind also, a certain amount of artistic culture may be superinduced; its powers of acquisition may be extended so far. But the spontaneous, exuberant, imaginative flow, the artistic nature at the core, never was, cannot be, included in the same individual. Aristotle could not be also a tragic poet; nor Newton a third-rate portrait-painter. The cost of one of the two modes of intellectual greatness is all that can be borne by the most largely-endowed personality; any appearances to the contrary are hollow and delusive.

Other instances could be given. Great activity and great sensibility are extreme phases, each using a large amount of power, and therefore scarcely to be coupled in the same system. The active, energetic man, loving activity for its own sake, moving in every direction, wants the delicate circumspection of another man who does not love activity for its own sake, but is energetic only at the spur of his special ends.

And once more. Great intellect as a whole is not readily united with a large emotional nature. The incompatibility is best seen by inquiring whether men of overflowing sociability are deep and original thinkers, great discoverers, accurate inquirers, great organizers in affairs; or whether their greatness is not limited to the spheres where feeling performs a part—poetry, eloquence, and social ascendency.

THE END.

INDEX.

THE END.

www.ingramcontent.com/pod-product-compliance
Lightning Source LLC
LaVergne TN
LVHW010255110826
845151LV00004B/1477

* 9 7 8 1 4 2 5 5 2 2 2 4 7 *